CHINA LITERATURE AND ART FOUNDATION
中国文学艺术基金会　资助项目
中国文学艺术发展专项基金

中国重要农业文化遗产影像志

A PHOTOGRAPHIC RECORD OF CHINA-NIAHS

高 扬 主编

EDITED BY GAO YANG

中国摄影出版传媒有限责任公司

China Photographic Publishing & Media Co., Ltd.

中国摄影出版社

目 录
CONTENTS

前言：新的遗产类型，新的保护范式   8
PREFACE：New Type of Heritage, New Paradigm of Conservation

华北地区 / North China   28

天津市 / Tianjin Municipality
皇家枣园的代表——天津滨海崔庄古冬枣园   30
A Representative of Royal Jujube Orchard—Cuizhuang Ancient Winter Jujube Orchard

河北省 / Hebei Province
传统漏斗架葡萄栽培体系——河北宣化城市传统葡萄园   38
Traditional Funnel Rack Viticulture System—Urban Agricultural Heritage of Xuanhua Grape Gardens
北方林粮间作典型模式——河北宽城传统板栗栽培系统   48
Typical Intercropping Mode of Forest and Crops in North China—Kuancheng Traditional Chestnut Cultivation System
北方梯田与石头文化的集成——河北涉县旱作梯田系统   56
Integration of Terraced Fields and Stone Culture in North China—Shexian Dryland Farming Terraces System

内蒙古自治区 / Inner Mongolia Autonomous Region
世界旱作农业源头——内蒙古敖汉旱作农业系统   66
The Birthplace of the World's Dryland Farming—Aohan Dryland Farming System
蒙古族游牧文化的缩影——内蒙古阿鲁科尔沁草原游牧系统   78
The Epitome of Mongolian Nomadic Culture—Aru Horqin Grassland Nomadic System

东北地区 / Northeast China   **88**

  辽宁省 / Liaoning Province

  南果梨母株所在地——辽宁鞍山南果梨栽培系统   90
  The Location of the Mother Plant of Nanguo Pear—Anshan Nanguo Pear Cultivation System

  传统林参共作模式——辽宁宽甸柱参传统栽培体系   98
  The Traditional Forest-ginseng Integrated Model—Kuandian Pillar Ginseng Cultivation System

华东地区 / East China   **106**

  江苏省 / Jiangsu Province

  沼泽洼地土地利用模式——江苏兴化垛田传统农业系统   108
  The Land Use Patterns in Swamp and Lowland Areas—Xinghua Duotian Traditional Agrosystem

  浙江省 / Zhejiang Province

  传统稻鱼共生农业生产模式——浙江青田稻鱼共生系统   120
  A Traditional Rice-fish Symbiotic Agricultural Production Model—Qingtian Rice-Fish Culture System

  陡坡山地高效农林生产体系——浙江绍兴会稽山古香榧群   130
  The High-efficiency Agroforestry Production System Applied for Slopes—Shaoxing Kuaijishan Ancient Chinese Torreya

  传统茶禅文化代表——浙江杭州西湖龙井茶文化系统   138
  A Representative of Traditional Tea-zen Culture—Hangzhou West Lake Longjing Tea Culture System

  低洼地区杰出的生态农业模式——浙江湖州桑基鱼塘系统   146
  An Outstanding Ecological Agriculture Model in Low-lying Areas—Huzhou Mulberry-dyke & Fish-pond System

  世界香菇起源地——浙江庆元香菇文化系统   156
  The Origin of the World's Shiitake Mushroom—Qingyuan Shiitake Mushroom Culture System

福建省 / Fujian Province

湿地山地立体农业生产体系——福建福州茉莉花与茶文化系统　　170
The Stereoscopic Agricultural Production System in Wetlands and Mountainous Areas—Fuzhou Jasmine and Tea Culture System

竹林、村庄、梯田、水系综合利用模式——福建尤溪联合梯田　　178
A Comprehensive Utilization Model of Bamboo Forests, Villages, Terraces and Rivers System—Youxi Lianhe Terraces

乌龙茶发源地——福建安溪铁观音茶文化系统　　188
The Birthplace of Oolong Tea—Anxi Tie Guanyin Tea Culture System

江西省 / Jiangxi Province

世界人工栽培稻源头——江西万年稻作文化系统　　196
The Manual Rice Cultivation Place in the World—Wannian Traditional Rice Culture System

最大的客家梯田——江西崇义客家梯田系统　　206
The Largest Hakka Terraces—Chongyi Hakka Terraces

山东省 / Shandong Province

沙地生态治理与经济发展的典范——山东夏津黄河故道古桑树群　　214
A Representative of Ecological Management and Economic Development in Sandy Land—Traditional Mulberry System in Xiajin's Ancient Yellow River Course

华中地区 / Central China　　224

湖北省 / Hubei Province

砖茶之源——湖北羊楼洞砖茶文化系统　　226
The Origin of Brick Tea—Yangloudong Brick Tea Culture System

湖南省 / Hunan Province

南方稻作文化与苗瑶山地渔猎文化融合体系——湖南新化紫鹊界梯田　　236
The Fusion System of Southern Rice Culture and Miao-Yao Montanic Fishing & Hunting Culture—Xinhua Ziquejie Terraces

世界原始稻作文化的"活化石"——湖南新晃侗藏红米种植系统   248

The "Living Fossil" of the World's Original Rice Culture—Xinhuang Red Rice Planting System of Dong Nationality

华南地区 / South China   256

广东省 / Guangdong Province

岭南茶文化代表——广东潮安凤凰单枞茶文化系统   258

A Representative of Lingnan Tea Culture—Chao'an Phoenix Single Cluster Tea Culture System

广西壮族自治区 / Guangxi Zhuang Autonomous Region

壮瑶人民农耕文明的结晶——广西龙胜龙脊梯田农业系统   268

The Crystallization of Farming Civilization of Zhuang and Yao Nationalities—Longsheng Longji Terraces

西南地区 / Southwest China   276

四川省 / Sichuan Province

川西北山地药用植物文化的代表——四川江油辛夷花传统栽培体系   278

A Representative of Typical Alpine Traditional Chinese Medicine in Northwest Sichuan—Jiangyou Traditional Magnolia Cultivation System

云南省 / Yunnan Province

山区稻作梯田的典型代表——云南红河哈尼稻作梯田系统   286

The Perfect Example of Mountainous Rice Terraces—Honghe Hani Rice Terraces System

世界茶树原产地和茶马古道起点——云南普洱古茶园与茶文化系统   298

The Origin of Tea Planting and Starting Point of Ancient Tea Horse Road—Pu'er Traditional Tea Agrosystem

传统核桃与农作物套作农耕模式——云南漾濞核桃—作物复合系统   310

The Traditional Intercropping Pattern of Walnut and Crops—Yangbi Walnut-Crop Complex System

云南壮族稻作文化的代表——云南广南八宝稻作生态系统   318

A Representative of Rice Culture of Yunnan Zhuang Nationality—Guangnan Babao Rice Farming Ecosystem

3000年水旱轮作的"活化石"——云南剑川稻麦复种系统   326

The "Living Fossil" of 3000-year-old Crop Rotation—Jianchuan Rice-Wheat Multiple Cropping System

贵州省 / Guizhou Province

传统稻鱼鸭共生农业生产模式——贵州从江侗乡稻鱼鸭系统　　334
Traditional Rice-fish-duck Symbiotic Agricultural Production Mode—Congjiang Dong's Rice-Fish-Duck System

西北地区 / Northwest China　　344

陕西省 / Shaanxi Province

干旱地区山地高效农林生产体系——陕西佳县古枣园　　346
High-efficiency Agroforestry Production System in Arid Areas—Jiaxian Traditional Chinese Jujube Garden

甘肃省 / Gansu Province

古梨树存量最多的梨树栽培体系——甘肃皋兰什川古梨园　　354
Pear Tree Cultivation System with the most Stock of Ancient Pear Trees—Gaolan Shichuan Ancient Pear Orchard

农、林、牧循环复合生产体系——甘肃迭部扎尕那农林牧复合系统　　362
Cyclic Agriculture-Forestry-Animal Husbandry Composite System—Diebu Zagana Agriculture-Forestry-Animal Husbandry Composite System

"千年药乡"的杰出代表——甘肃岷县当归种植系统　　370
An Outstanding Representative of Millennium Medicine Township—Minxian Angelica Planting System

宁夏回族自治区 / Ningxia Hui Autonomous Region

独特环境、独特品种、独特技艺——宁夏灵武长枣种植系统　　378
Unique Environment, Variety and Artistry—Lingwu Long Jujube Planting System

新疆维吾尔自治区 / Xinjiang Uygur Autonomous Region

大型地下农业水利灌溉工程——新疆吐鲁番坎儿井农业系统　　386
Large-Scale Underground Agricultural Irrigation Project—Turpan Karez Agricultural System

哈密地域文化与财富的标志——新疆哈密市哈密瓜栽培与贡瓜文化系统　　394
Symbol of Regional Culture and Wealth of Hami City—Hami Melon Cultivation and Tribute Melon Culture System

附录 / APPENDIX 402

中国重要农业文化遗产名单 402
List of China Nationally Important Agricultural Heritage Systems

中国的全球重要农业文化遗产名单 408
List of Globally Important Agricultural Heritage Systems of China

注：本书所涉遗产包括了原农业部发布的第一、二批中国重要农业文化遗产，共39项。

Note: This book covers the first and second batches of 39 China-NIAH sites released by Ministry of Agriculture, which is called Ministry of Agriculture and Rural Affairs now.

前　言

新的遗产类型，新的保护范式

目前所谈的农业文化遗产，主要指的是联合国粮农组织认定的"全球重要农业文化遗产"（GIAHS）和中国农业农村部（原农业部）认定的"中国重要农业文化遗产"（China-NIAHS）。

一、农业文化遗产：一种新的遗产类型

按照2015年发布的《重要农业文化遗产管理办法》，重要农业文化遗产是指我国人民在与所处环境长期协同发展中世代传承并具有丰富的农业生物多样性、完善的传统知识与技术体系、独特的生态与文化景观的农业生产系统，包括由联合国粮农组织认定的全球重要农业文化遗产和由农业农村部认定的中国重要农业文化遗产。显然，重要农业文化遗产（以下简称为农业文化遗产）是一类特殊的遗产[1]，之所以说"特殊"，是因为它除了具有一般"遗产"的历史特征外，至少还需要从经济、生态、技术、文化、景观等维度进行理解，而这也正是联合国粮农组织全球重要农业文化遗产和农业农村部中国重要农业文化遗产认定的五个核心标准。

农业文化遗产概念提出后，受到的质疑主要来自于三个方面：一是农业文化遗产属于农业历史还是农业遗产范畴；二是农业文化遗产属于文物还是文化遗产范畴；三是农业文化遗产是否属于文化景观范畴。农业文化遗产与这些概念有着密切的联系，但又有所区别，正是因为这些原因，才使农业文化遗产成为一类新的遗产。

农业文化遗产是一类特殊的农业历史或农业遗产。中国农业的起源，可以追溯到距今一万多年的史前年代。农业不仅开启了人类文明之源，而且凝聚了人类智慧，形成了一部内容极其丰富的农业史。在古代，还没有形成"农业史学"的学术概念，但是以农业为对象的学问是与史俱来、与史俱存的，历朝历代都有人在做农业史的学问。20世纪20年代末，金陵大学、中央大学相继开设"中国农业史"课程，中国出现了一批筚路蓝缕的农业史学家，如《中国田制史》作者万国鼎、《中国水利史》作者郑肇经、《中国救灾史》作者邓云特、《中

1　闵庆文. 全球重要农业文化遗产——一种新的世界遗产类型. 资源科学，2006，28（4）：206-208

国渔业史》作者李士豪等，他们的开创性工作打下了农业史的学科基础[2]。

"农业遗产"并不同于一般意义上的"农业历史"。农业历史的研究重点在于对农业发生发展和历史演变的探索和陈述，而农业遗产则是特指农业史演进过程中所遗存下来的具有继承价值的优良部分。概而言之，农业史学侧重于对历史过程的客观陈述，农业遗产侧重于对传统农业要素的价值判断[3]。

不过，早期的农业遗产研究，更多的仍然是农业历史的研究，从20世纪50年代有关高校的相关研究机构名称即可窥知。例如，在西北农学院称为"古农学研究室"，在北京农业大学称为"农业史研究室"，在华南农学院则称为"农业历史遗产研究室"，只有1955年7月成立的由中国农业科学院和南京农学院双重领导的中国农业遗产研究室，覆盖研究范围相对宽泛。在新中国成立后的前17年，农业遗产的整理重点是古农书整理校注和传统文献中农业史资料的搜集，这方面的工作很有成效，为农史学科的发展奠定了坚实的基础。这种情况也容易使人产生错觉，似乎"农业遗产"就是指古农书和有关文献[4]。

从"农业遗产"到"农业文化遗产"，则又是一个巨大的变化。南京农业大学王思明等认为，从原来静止不变的农业遗产资料的研究向活态、原生态农业遗产研究和保护的转变，是近年来中国农业遗产研究的重要变化[5]。

但从研究目的、研究对象、研究方法等方面分析，农业文化遗产与一般意义的农业遗产则有着很大不同。在我国，"农业文化遗产"可能最早出现于曹幸穗研究员于2004年3月5日以全国政协委员身份提出的《关于将中国农业文化遗产纳入中国民族民间文化保护工程的提案》中。后来这个提案被转到国家文化部（现文化和旅游部）答复办理。文化部领导非常重视，及时与提案人取得联系，并批复在中国农业博物馆启动"中国农业文化遗产保护项目"，于2004年4月至2006年6月实施了"贵州从江县和威宁县的农业文化保护试点"[6]。"农业文化遗产"最早出现于学术期刊，可能是闵庆文研究员于2006年发表在《资源科学》上的《全球重要农业文化遗产——一种新的世界遗产类型》一文。文章重点介绍了联合国粮农组织提出的"全球重要农业文化遗产"（GIAHS）的概念和所推进的项目。

农业历史、农业遗产、农业文化遗产有着内在的联系，但又有着显著的区别。时至今日，作为国家一级学会的中国农业历史学会，以及《中国农史》等学术期刊，依然在推动着农史学科的发展。在农业遗产领域，中国人继承了农业大国和农业古国的传统，保持了引领世界潮流的领跑地位[7]。而重点关注农业遗产中"活态"部分的联合国粮农组织的"全球重要农业文化遗产"和中国农业农村部的"中国重要农业文化遗产"发掘与保护的研究和实践，则为农史学科的发展提供了新的契机和动力[8]。

农业文化遗产是一类特殊的文物或文化遗产。文物是人类在社会活动中遗留下来的具有历史、艺术、科学价值的遗物和

2　曹幸穗. 农业文化遗产相关概念的辨析. 遗产与保护研究, 2016, 1（1）:90-94
3　曹幸穗. 农业文化遗产相关概念的辨析. 遗产与保护研究, 2016, 1（1）:90-94
4　李根蟠. 农史学科发展与"农业遗产"概念的演进. 中国农史, 2011,（3）:121-128
5　王思明, 卢勇. 中国的农业遗产研究：进展与变化. 中国农史, 2010,（1）:3-11
6　曹幸穗. 农业文化遗产相关概念的辨析. 遗产与保护研究, 2016, 1（1）:90-94
7　曹幸穗. 农业文化遗产相关概念的辨析. 遗产与保护研究, 2016, 1（1）:90-94
8　李根蟠. 农史学科发展与"农业遗产"概念的演进. 中国农史, 2011,（3）:121-128

遗迹[9]。从这个定义可知，文物是指具体的物质遗存，并具有两个基本特征：必须是由人类创造的，或者是与人类活动有关的；必须是已经成为历史的过去，不可能再重新创造的。

在国际社会，由联合国教科文组织（UNESCO，以下简称"教科文组织"）会议通过的一些有关保护文物的国际公约中，一般把文物称作"文化财产（Cultural Property）"或者"文化遗产（Cultural Heritage）"，二者所指的内容并不是等同的。从公约所列举的具体内容来看，前者是指可以移动的文物，后者是指不可移动的文物。

中国文化遗产研究院的演变可以反映出保护对象和保护思路上的一些变化。1935年，中华民国政府决定成立旧都文物整理委员会及其执行机构北平文物整理实施事务处；1949年新中国成立后，北平文物整理委员会（旧都文物整理委员会为其前身）及其工程处正式更名为北京文物整理委员会；后几经改变，于1973年6月成为文物保护科学技术研究所，1990年8月成为中国文物研究所，2007年8月更名为中国文化遗产研究院。

将传统的文物扩展为活态的文化遗产进行保护的时间并不长。2006年被国务院批准为第六批全国重点文物保护单位的，一个是位于河北省黄骅市齐家务乡娘娘河畔的聚馆村的聚馆古贡枣园，是中国现存最大最古老的古冬枣园，有数千年历史；另一个是与万里长城、京杭大运河并称为"中国古代三大工程"的新疆坎儿井。特别是聚馆古贡枣园的列入，开创了中国文物保护的新门类，成为最早的一处农业类、植物类全国重点文物保护单位。2013年，云南红河哈尼梯田、普洱景迈山古茶园入选第七批全国重点文物保护单位。此外，滨海新区崔庄古枣园、新化紫鹊界梯田、从江加榜梯田、兴化垛田等先后被列入天津、湖南、贵州、江苏省级文物保护单位。颇有意思的是，上述所列，除聚馆古贡枣园外，其他均为全球重要农业文化遗产或中国重要农业文化遗产项目。

因为"农业文化遗产"中的"文化"两字，使很多人简单地将农业文化遗产归属于文化遗产。2004年，曹幸穗教授以全国政协委员身份提出的《关于将中国农业文化遗产纳入中国民族民间文化保护工程的提案》，是由文化部答复办理的；2014年，全国政协"保护和利用好我国农业文化遗产"的调研活动，是由文史和学习委员会组织开展的；2016年，李文华院士牵头提出的《关于加强我国农业文化遗产研究与保护工作的建议》，是由负责文化部门工作的国家领导人批示的。2017年12月23日，受国务院委托，文化部部长雒树刚向十二届全国人大常委会第三十一次会议作国务院关于文化遗产工作情况的报告。在谈到"建立了多层级的文物和非物质文化遗产保护名录"工作时他指出："开展了中国重要农业文化遗产项目认定工作，农业部认定中国重要农业文化遗产项目91个。"

关于农业文化遗产的中文名称，最初存在一些争议，争议的一个重要方面在于联合国粮农组织项目名称只有"农业（Agriculture）"没有"文化（Culture）"的提法，闵庆文研究员曾对此进行了解释[10]。关于在英文中没有"Culture"一词，该项目的原总协调人、联合国粮农组织官员帕尔维兹（Parviz Koohafkan）先生认为，"Agriculture"本身就是一种"Culture"。李根蟠研究员也认为，农业也是文化的一种，故将"农业遗产"称为"农业文化遗产"亦无大碍[11]。曹幸穗研究员也表达

9　李晓东. 文物学. 北京：学苑出版社，2015
10　闵庆文. 关于"全球重要农业文化遗产"的中文名称及其他. 古今农业，2007, (3):116-120
11　李根蟠. 农史学科发展与"农业遗产"概念的演进. 中国农史，2011, (3):121-128

了类似的意见，暂且仍然特指性地将"Agricultural Heritage Systems"或"Agri-cultural Heritage"称为"农业文化遗产"，而不是学术上泛指的农业遗产或狭义的农业文化遗产[12]。

当然，在文化遗产中还有一类是非物质文化遗产。在非物质文化遗产中，最具典型性的农业类项目当属"二十四节气"。2006年，"二十四节气"被列入第一批国家级非物质文化遗产代表项目；2016年11月30日，被正式列入《世界非物质文化遗产名录》中，成为中国的第39个世界非物质文化遗产项目。

农业文化遗产是一类特殊的景观或文化景观。"景观（Landscape）"原是一个地理学名词，一般意义上是指一定区域呈现的景象，即视觉效果。这种视觉效果反映了土地及土地上的空间和物质所构成的综合体，是复杂的自然过程和人类活动在大地上的烙印。《中国大百科全书·地理学》概括了地理学中对景观的几种理解：某一区域的综合特征，包括自然、经济、文化诸方面；一般自然综合体；区域单位，相当于综合自然区划等级系统中最小的一级自然区；任何区域单位[13]。生态学上的景观是指由相互作用的拼块或生态系统组成，以相似的形式重复出现的一个空间异质性区域，是具有分类含义的自然综合体。

文化景观（Cultural Landscape）一词在20世纪20年代起就已普遍应用，它是人类在地表活动的产物，是自然风光、田野、建筑、村落、厂矿、城市、交通工具和道路以及人物和服饰等所构成的文化现象的复合体，反映文化体系的特征和一个地区的地理特征。美国地理学家C.O.索尔（Carl Ortwin Sauer）在1925年发表的专著《景观的形态》（*The Morphology of Landscape*）中主张用实际观察地面景色来研究地理特征，通过文化景观来研究文化地理。

文化景观引起更广泛关注则与联合国教科文组织对于文化遗产的概念和内涵的拓展有关，1992年增加了"文化景观"这一新的类型（也有人认为文化景观仍属于文化遗产的范畴）。在所定义的三种文化景观类型中，与农业密切相关的是"有机进化的景观"，它产生于最初始的一种社会、经济、行政以及宗教需要，并通过与周围自然环境相联系或相适应而发展到目前的形式。特别是其中的一种次类别"持续性景观"，指在当地与传统生活方式相联系的社会中保持一种积极的社会作用，而且其自身演变过程仍在进行之中同时又展示了历史上其演变发展的物证。

在联合国教科文组织的世界文化遗产中，虽然没有"农业遗产"这一类型，但是有许多农业类遗产被列入"文化景观"的范围。例如，以苏巴克灌溉系统为核心的印度尼西亚巴厘文化景观、包括大量农耕文化元素的法国卢瓦尔河谷、位于喀斯和塞文的地中海农牧文化景观、包括葡萄园景观在内的勃艮第与香槟葡萄园、瑞典奥兰南部农业景观、古巴东南最早的咖啡种植园等。

比较有意思的是，目前有几个项目获得了联合国粮农组织与教科文组织的分别认定。比较有代表性的是：菲律宾科迪勒拉山区稻作梯田1995年被认定为世界文化遗产，依富高稻作梯田系统2005年被认定为全球重要农业文化遗产；我国云南红河哈尼稻作梯田系统2010年被认定为全球重要农业文化遗产，2013年哈尼稻作梯田被认定为世界文化遗产；阿拉伯联合酋长国阿尔恩文化遗址（包括绿洲地区）2011年被认定为世界文化遗产，艾尔与里瓦绿洲传统椰枣种植系统2015年被认定为全球重要农业文化遗产；伊朗喀山坎儿井灌溉系统2014年

---

[12] 曹幸穗. 农业文化遗产相关概念的辨析. 遗产与保护研究，2016，1（1）：90-94
[13] 中国大百科全书出版社编辑部. 中国大百科全书·地理学. 北京：中国大百科全书出版社，1990

被认定为全球重要农业文化遗产，坎儿井 2016 年被认定为世界文化遗产。

二、农业文化遗产保护：需要建立新的范式

虽然经过 10 多年的努力，农业文化遗产的发掘与保护已经取得了很大进展，并产生了良好的生态、社会与经济效益，但与其他类型的自然与文化遗产相比，农业文化遗产仍然不被人所熟知。相比人们耳熟能详的神农架、梵净山等自然遗产，长城、故宫等文化遗产，泰山、黄山等混合遗产，昆曲、武术等非物质文化遗产，农业文化遗产还属于"藏在深闺人未识"，农业文化遗产发掘与保护工作任重道远。

从前面关于农业文化遗产概念及所被认定的典型项目中不难发现，农业文化遗产有其自身特点，而这些特点决定了不能沿用一般意义上的文物或文化遗产保护思路，需要建立新的保护范式。

农业文化遗产的主要特点。一是"活态性"，即农业文化遗产是有人参与、至今仍在使用、具有较强的生产与生态功能的农业生产系统，系统的直接生产产品和间接生态与文化服务依然是农民生计保障和乡村和谐发展的重要基础；二是"动态性"，即随着社会经济发展与技术进步以及满足人类不断增长的生存与发展需要，传统农业生产系统不断发生着变化，但这种变化是系统稳定基础上的结构与功能的调整；三是"适应性"，即随着不同地区或不同历史时期自然条件的变化，传统农业生产系统不断发生着适应性的变化，这种变化是系统稳定基础上的协同进化，充分体现出人与自然和谐的生存智慧；四是"复合性"，即农业文化遗产不仅包括一般意义上的传统农业知识和技术，还包括那些历史悠久、结构合理的传统农业景观，以及独特的农业生物资源与丰富的生物多样性，体现了自然遗产、文化遗产、文化景观、非物质文化遗产的复合特点；五是"战略性"，即农业文化遗产对于应对全球化和全球变化，保护生物多样性，保障生态安全与粮食安全，有效缓解贫困，促进农业可持续发展和乡村振兴等，均具有重要的战略意义；六是"多功能性"，即农业文化遗产具有多样化的物质性生产功能和巨大的生态与文化价值，充分体现出食品保障、原料供给、就业增收、生态保护、观光休闲、文化传承、科学研究等多种功能；七是"可持续性"，即历经千百年传承至今的农业文化遗产，是一类典型的"社会—经济—自然"复合生态系统，具有结构合理的生态系统、多样产出的经济系统与和谐稳定的社会系统，是人地和谐、可持续发展的典范；八是"濒危性"，主要是由于政策与技术原因以及比较效益等，使传统农业系统面临着威胁，主要表现为农业生物多样性的减少和丧失、传统农业技术和知识体系的消失以及农业生态系统结构与功能的破坏[14]。

农业文化遗产保护的基本原则。农业文化遗产是以农业生产功能为基础的系统性遗产，保护的基本原则是"在发掘中保护，在利用中传承"，具体就是"保护优先、适度利用，整体保护、协调发展，动态保护、适应管理，活态保护、功能拓展，现地保护、示范推广，多方参与、惠益共享"，其核心是"整体、动态、活化"，即整体性保护、动态性管理、活化性利用。

随着社会经济的发展和人们生活水平的提高，农业生产也在不断进步，农业生产系统的结构、要素、功能也将发生相应

14　闵庆文. 农业文化遗产的概念特点以及保护与发展. 农民日报，2013 年 2 月 8 日第 4 版

的变化，农业文化遗产很难真正做到"原汁原味"。而且，农业文化遗产多处于经济落后、生态脆弱、文化底蕴丰厚的地区，肩负着经济发展、生态保护、文化传承的多重任务。过分强调"原汁原味"的保护而忽视了区域发展，难以调动当地居民遗产保护的积极性，难以实现保护的目的，需要探索动态保护与适应性管理的新思路。

"变"与"不变"是农业文化遗产保护需要关注的重要方面。"变"是绝对的，"不变"是相对的，关键是什么可以变（农业文化遗产系统的非关键性要素可以改变）、什么不可以变（农业文化遗产系统的关键要素不可以改变），以及变的"度"如何把握。所谓动态保护，就是不应当过分强调"原汁原味"或者"一成不变"的"冷冻式保存"，而是应当根据实际情况进行适当的调整，实现农业文化遗产系统的"活态传承"与"持续发展"，但系统的基本结构与功能及重要的物种资源、农业景观、水土资源管理技术等不应发生改变，与之相关的民族文化与传统知识也不应发生大的改变[15]。没有保护，就失去了发展的基础，则无以谈发展；但没有发展，则可能会失去保护的动力，难以实现保护的目的。

农业文化遗产保护的关键机制。一是需要建立农业文化遗产保护的"政策激励机制"。农业文化遗产蕴含着对于当今和未来农业发展具有重要价值的生物、文化和技术"基因"：传统农业生产系统中的许多重要动植物遗传资源及相关的生物多样性，在维持生态系统稳定和服务功能发挥等方面具有重要作用；农业生产过程中创造的诸如侗族大歌、哈尼四季生产调、青田鱼灯舞等丰富多彩的歌舞以及民俗、饮食、建筑等物质与非物质文化遗产，对于农耕文化传承、农村社会和谐等具有重要意义；稻田养鱼、桑基鱼塘、农林复合、梯田耕作、间作套种等传统农业生产技术，对于现代生态循环农业发展具有借鉴意义。因此，应当研究并实施农业文化遗产的生态与文化保护补偿，即参照对于自然生态保护的思路和做法，对农业生物多样性与农业生态景观进行生态保护补偿；参照对于文物、非物质文化遗产和传统村落保护的思路和做法，对农业技术与文化多样性进行文化保护补偿。

二是需要建立农业文化遗产保护的"产业促进机制"。农业文化遗产除了具有直接的生产功能外，还具有重要的生态功能和文化功能，这为拓展农业多重功能和促进农业提质增效、农民就业增收、农村和谐稳定奠定了资源基础。农业文化遗产地具有发展"第六产业"的先天优势。特有的农业物种与生物资源、相对丰富的劳动力资源，以及传统的文化习俗和优美的乡村景观，成为发展劳动密集型的特色农业和农产品加工业、手工艺品制作、生态与文化旅游以及生物资源产业、文化创意产业等的基础。应当在坚持农业生产的前提下，积极发展有文化内涵的生态农产品以及农产品加工业、食品加工业、生物资源产业、文化创意产业、乡村旅游产业等为主要内容的多功能农业，逐步建立起农业功能拓展、"三产"融合发展的新型农业产业模式，实现农民从"农业生产者"向"多种经营者"的转变，农事活动、乡村景观、传统民俗、生态环境向生态与文化旅游资源的转变，原来自给自足的农产品向具有更高附加值的特色农产品、高端消费品和旅游纪念品的转变。

三是需要建立农业文化遗产保护的"多方参与机制"。农业文化遗产是先民创造、世代传承并不断发展的传统农业生产系统，其所有者应当是依然从事农业生产的"农民"，他们理

---

15 闵庆文."五位一体"的动态保护——谁来保护？如何保护？. 中华儿女，2016,(1):42-44

应成为农业文化遗产的最主要的保护者,同时也应当是农业文化遗产保护的最主要的受益者。但必须看到,之所以要对农业文化遗产进行保护,正是因为它们在现今条件下面临着威胁,不具有竞争力而处于"濒危"状态。如果仅靠农民进行保护,不仅难以实现保护的目的,而且把属于全人类共有共享的"遗产"保护重任压到弱势群体身上,也是不公平的。

为此,应当建立政府推动、科技驱动、企业带动、社区主动、社会联动的"五位一体"的多方参与机制。由政府发挥主导作用,制定相关保障性政策,实施规范化管理,组织规划编制和实施,负责资金筹措等,将农业文化遗产保护与利用纳入地方发展的总体布局中。充分发挥科技的支撑作用,组织农业生态、农业历史、农业文化、农业经济、农村发展等领域的专家,发掘、评估农业文化遗产的价值,分析农业文化遗产系统可持续发展机制,协助编制科学、可操作的保护与发展规划,进行传统知识与经验的理论提升。注重发挥企业在农业文化遗产保护与发展中的特殊作用,有效提高产品开发、市场开拓、资金投入、产业管理等方面的水平。充分调动农村社区居民保护农业文化遗产的主动性,提高他们的自信心,让他们真正成为农业文化遗产保护成果的最主要受益者。提高公众的参与意识,营造良好的社会氛围,探索社区支持农业、认养制度、志愿者制度等新模式[16]。

三、影像志:农业文化遗产保护的有益探索

影像志对于自然与文化遗产保护具有重要作用。近年来,通过影像技术手段保护自然与文化遗产已经引起重视。例如,为了促进保护和传承非物质文化遗产,国务院办公厅在2005年3月26日发布的《关于加强中国非物质文化遗产保护工作的意见》中强调,"要运用文字、录音、录像、数字化多媒体等各种方式,对非物质文化遗产进行真实、系统和全面的记录,建立档案和数据库。"

围绕影像之于文化遗产的保护,有不少学者或学子展开过研究和论述。崔莹认为,在"非遗"保护语境体系中,影像因强大的"视听结合"功能显然成为以传播为手段进而实现"非遗"间接保护功能的传播者。可以把这种传播方式看作是"仪式"与影像同构,即在特定文化空间中呈现"仪式"表征,同时又可借鉴影像深描的表述范式与编码形成紧密结合,使得影像置身于不同的"场域",以一个更自由的角色与"非遗"进行融合,传播经过精炼和改造的文化元素或地域文化的代表符号[17]。林礼顺关于用影像对非物质文化遗产进行保护的多种优越性的阐述具有借鉴意义:首先,逼真直观的影像更便于受众了解非物质文化遗产的核心内容,也可以更好地保护"非遗"的原生态;其次,影像有助于将无形的非物质文化遗产转变为可以长久保存的物质形态;再次,影像化能够将非物质文化遗产中大部分的文化事象转变为影像的表述方式,涵盖面较广,同时还可以更为完整深入地记录这些非物质文化遗产的社会文化语境,具有全方位与立体型的直观阐述体系;第四,影视化表述的传播渠道较广,既可以以故事片、纪录片的形式传播,又可以作为影像档案长久留存;第五,影像化的表述方式能够为地方化文化传统的重建与再发展提供影像化的依据,为非物质文化遗产的活态保护提供影像化基础。从认同的角度来看,影像化对非

---

16 闵庆文."五位一体"的动态保护——谁来保护?如何保护?.中华儿女,2016,(1):42-44;又见闵庆文.农业文化遗产保护的关键机制.光明日报,2016年8月19日10版
17 崔莹.论影像化在非物质文化遗产保护中的作用和意义.云南民族大学学报(哲学社会科学版),2018,35(6):38-42

物质文化遗产的保护与大面积传播不仅能够强化民众的集体认同感，还能影响民众的文化与社会认同，为我们的后代对非遗拥有良好的集体记忆打下坚实的基础，这对于时代的快速变迁无疑是一种良性的适应，使他们一方面可以增加更多的自豪感，另一方面也会更加自愿地保护非遗，并对中国传统文化自觉地加以传承与再发展。而经由影视媒介的象征符号与意义体系的影视化表达，中国人的文化认同与族群身份将得以强化[18]。

静态的照片虽然不具有动态影像那样的功能，但在遗产保护中的档案记录，这些照片广泛传播，易于读者接受，有利于增加文化认同、提高文化自信，作用同样不可小觑。2014年中国摄影出版社组织编辑出版的《中国世界遗产影像志》，就是一个很好的例证。该书用图文并茂的方式，多角度纵深地展示了中国世界遗产的独特魅力。几位专家的评价就是最好的说明：中国科学院院士刘嘉麒认为，"本书荟萃了我国自然与文化的精华，是文化艺术的经典"。历史地理学家葛剑雄告诉读者，"如你一时去不了，可以此书神游。如你准备去，可持此书导游"。世界遗产影像专家周梅生则认为，该书是"世界遗产的影像证明，中华文明的史志图鉴，人与自然的视觉乐章"[19]。

利用影像志方式保护农业文化遗产具有独特作用。因为农业文化遗产是具有生产功能的活态的传统农业系统，实行系统性保护应为首选，即通过重要农业文化遗产地的认定，划定核心保护区域，确定关键保护要素，采取有效保护措施，变单一的农业生产为多功能农业发展，让农民通过参与保护而获得收益，让农业通过提高效益而持续发展，让遗产通过持续发展而传承弘扬。但因为农业文化遗产是动态演变的系统，某些要素的改变难以避免，对那些可能发生改变的要素或组分采用（生态）博物馆或（物种）资源库进行保存、利用影像及信息手段进行记录、利用建立传承基地等方法进行传承就变得很有必要。

因此，对于农业文化遗产而言，用影像记录保护是一种方式，尽管可能不是最好的，更不可能是唯一的方式，但却可以发挥其独特作用。一是可以起到"保存"的作用，通过图像档案的方式记录下那些可能难以通过活态方式留存的某些要素与景观，发挥其文献与学术价值；二是可以克服农业生产与农业景观的季节性的弊端，让人在同一时间了解不同时期、不同季节的景象；三是可以发挥科学传播的作用，让更多的人更加形象地了解农业文化遗产，特别是让那些没有机会深入实地的人在一定程度上感受农业文化遗产的魅力；四是可以起到科普教育的作用，让更多的人透过影像感悟农业文化遗产所蕴含的丰富内涵。

以中英文对照形式出版"中国重要农业文化遗产影像志"将有助于推动中国优秀农耕文化走向世界。中国是世界农业重要起源地之一，有着上万年农耕历史。农耕文明不仅是中国古代文明的根基，而且对世界农业文明的发展也产生了十分深远的影响。

农业文化遗产发掘与保护目前已经取得了较为广泛的国际共识。针对现代农业发展所带来和所面临的资源大量消耗、生态系统退化、生物多样性丧失、环境污染严重、传统技术丢失、传统文化传承受阻等问题，2002年，联合国粮农组织（FAO）在全球环境基金（GEF）支持下，联合有关国际组织和国家，发起全球重要农业文化遗产（GIAHS）保护倡议，旨在建立全球重要农业文化遗产及其有关的景观、生物多样性、知识和文化保护体系，并在世界范围内得到认可与保护，使之成为可持

18　林礼顺. 非物质文化遗产影像化保护的方式. 东南传播，2014，(10)：54-56
19　赵迎新主编. 中国世界遗产影像志. 北京：中国摄影出版社，2014

续管理的基础。该倡议将努力促进地区和全球范围内对当地农民和少数民族关于自然和环境的传统知识和管理经验的更好认识，并运用这些知识和经验来应对当代发展所面临的挑战，特别是促进可持续农业的振兴和农村发展目标的实现。自2005年在6个国家选择了5个不同类型的传统农业系统作为首批保护试点以来，经过10多年的努力，现已有21个国家的57个项目被列入全球重要农业文化遗产名录。

中国是联合国粮农组织全球重要农业文化遗产倡议的最早响应者、积极参与者、坚定支持者、成功实践者、重要推动者和主要贡献者。略举几例即可说明：2005年成功将浙江青田稻鱼共生系统推荐为首批试点并第一个正式授牌，截至目前已有15个项目得到联合国粮农组织认定，数量居各国之首；2012年率先开展国家级农业文化遗产发掘工作，农业农村部已发布4批91项中国重要农业文化遗产；2013年发起成立东亚地区农业文化遗产研究会（ERAHS），并率先承办区域性学术研讨活动；2014年和2016年成功推动将GIAHS写入《亚太经合组织（APEC）粮食安全部长会议宣言》和《二十国集团（G20）农业部长会议宣言》；2014年起连续举办"GIAHS高级别培训班"，已培养来自60多个国家和国际组织的100多位学员，浙江青田、绍兴与湖州、广西龙胜、山东夏津、江苏兴化、河北宣化、福建福州、尤溪等重要农业文化遗产地已成为重要的培训教学基地；2014年成立中国农学会农业文化遗产分会，并连续举办年度学术交流与经验分享；2015年颁布《重要农业文化遗产管理办法》，使我国成为第一个颁布管理办法的国家，同年开始GIAHS监测评估工作；2016年开展全国性农业文化遗产普查，并于当年发布408项具有潜在保护价值的农业文化遗产。此外，李文华院士2011年首任GIAHS项目指导委员会（ST）主席；闵庆文研究员于2013年获得联合国粮农组织第一个"全球重要农业文化遗产特别贡献奖"；金岳品先生于2014年因农业文化遗产保护而获得"亚太地区模范农民"称号；闵庆文研究员2016年首任GIAHS科学咨询小组（SAG）主席；等等。

从某种意义上说，中国的农业文化遗产正在影响世界农业可持续发展和乡村振兴，中国的农业文化遗产发掘与保护经验已经并将继续引领国际农业文化遗产保护运动。而本书的出版，无疑将有助于进一步推进中华优秀农耕文化走向世界。

农业文化遗产很重要，这句话可能是对其重要意义最好的说明："农耕文化是我国农业的宝贵财富，是中华文化的重要组成部分，不仅不能丢，而且要不断发扬光大。"

让我们记住联合国粮农组织的口号：农业文化遗产不是关于过去的遗产，而是关乎人类未来的遗产。

闵庆文

2019年4月2日

（作者为全国政协委员，联合国粮农组织全球重要农业文化遗产科学咨询小组副主席，农业农村部全球/中国重要农业文化遗产专家委员会副主任委员兼秘书长，中国生态学学会副理事长，中国农业历史学会副理事长，中国农学会农业文化遗产分会副主任委员兼秘书长，中国科学院地理科学与资源研究所研究员、自然与文化遗产研究中心副主任）

PREFACE

New Type of Heritage, New Paradigm of Conservation

At present, the agri-cultural heritage, a new type of heritage, mainly refers to the "Globally Important Agricultural Heritage Systems (GIAHS)" identified by the Food and Agriculture Organization of the United Nations (FAO) and the "China Nationally Important Agricultural Heritage Systems (China-NIAHS)" identified by the Ministry of Agriculture and Rural Affairs of the People's Republic of China (formerly the Ministry of Agriculture of the People's Republic of China).

1. Agri-cultural Heritage is a New Type of Heritage

In accordance with the *Measures for the Management of Important Agricultural Heritage Systems* issued in 2015, the Important Agricultural Heritage Systems refer to the agricultural production systems inherited from generation to generation by the Chinese people in their long-term coordinated development with the environment, featured by rich agricultural biodiversity, sound traditional knowledge and technology system, unique ecological and cultural landscapes, including the GIAHS identified by FAO and the China-NIAHS identified by the Ministry of Agriculture and Rural Affairs. Obviously, the Important Agricultural Heritage Systems (hereinafter referred to as agri-cultural heritage) are a special kind of heritage.[1] They are special because they need to be understood at least from the economic, ecological, technological, cultural, landscape and other perspectives besides the general historical characteristics of "heritage". The afore-mentioned perspectives are also the five core standards for the identification of FAO's GIAHS and China-NIAHS of the Ministry of Agriculture and Rural Affairs.

After the proposal of the concept of agri-cultural heritage, it is mainly impugned in three aspects. First, agri-cultural heritage belongs to the category of agricultural history or agricultural heritage. Second, agricultural heritage systems belong to the category of cultural relics or cultural heritage. Third, agricultural heritage systems belong to the category of cultural landscape. Indeed agricultural heritage systems are closely related to these concepts, but differ from each other. It is precisely because of these reasons that make agricultural heritage systems a new type of heritage.

Agri-cultural heritage is a special type of agricultural history or

---

[1] Min Qingwen. *GIAHS: A New Kind of World Heritage*. Resources Science [J]. 2006, 28（4）:206-208.

agricultural heritage. The origin of agriculture in China can be traced back to the prehistoric times (more than 10,000 years ago). Agriculture not only initiated the source of human civilization, but also condensed human wisdom, forming the exquisite history of agriculture. In ancient times, the academic concept of "agricultural history" did not exist, but the knowledge of agriculture as research object was both simultaneous and coexistent with the history. The agricultural history has been studied in all dynasties of China. At the end of the 1920s, Jinling University and Central University offered the course of "Chinese Agricultural History" successively. A number of painstaking agricultural historians appeared in China, such as Wan Guoding, the author of *History of Field System in China*, Zheng Zhaojing, the author of *History of Water Conservancy in China*, Deng Yunte, the author of History of Disaster Relief in China and Li Shihao, the author of *History of Fishery in China*. Their pioneering work laid a foundation for the disciplinary basis of agricultural history[2].

"Agri-cultural heritage" is different from "agricultural history" in general sense. The research of agricultural history focuses on exploring and stating the occurrence and development of agriculture and its historical evolution, while the research of agricultural heritage focuses on the good parts with inheritance value that remains from the evolution of agricultural history. In general, agricultural history is oriented to the objective statement of historical process, while agricultural heritage is oriented to the value judgment of traditional agricultural elements.[3]

However, the early research on agricultural heritage is still more about agricultural history. It can be seen from the names of relevant research institutions in universities in the 1950s. For example, it is called the Paleo-Agricultural Institute in the Northwest College of Agriculture, the Agricultural History Institute in the Beijing Agricultural University, and the Agricultural History Heritage Institute in the South China Agricultural College. Only the Chinese Agricultural Heritage Institute which was established in July 1955 under the dual leadership of the Chinese Academy of Agricultural Sciences and the Nanjing Agricultural College covered a relatively wide range of research. During the first 17 years after the founding of the People's Republic of China, the key work of agricultural heritage review was the collation and annotation of ancient agricultural books and the collection of agricultural history materials in traditional documents. This work has been very effective and laid a solid foundation for the disciplinary development of agricultural history. This situation is easy to cause illusion. It seems that "agricultural heritage" definitely refers to ancient agricultural books and related documents.[4]

From "agricultural heritage" to "agri-cultural heritage" ("agricultural heritage systems") is another tremendous change. Wang Siming et al. of Nanjing Agricultural University believe that the transformation from the original static research of agricultural heritage data to the research and protection of living and original ecological agricultural heritage is an important change of China's research on agricultural heritage in recent years.[5]

However, from the analysis of research purposes, research objects and methods, agricultural heritage systems are quite different from agricultural heritage in general sense. In China, the term of "agricultural heritage systems" was probably first mentioned in the member of the National Committee of CPPCC Professor Cao Xingsui's *Proposal on Incorporating Chinese Agri-cultural Heritage into the Protection Project of Chinese Ethnic and Folk Culture* which was put forward on March 5, 2004. Later, this proposal was transferred to the Ministry of Culture (now the Ministry of Culture and Tourism) for dealing with. The leaders of the Ministry of Culture attached great importance to contacting the proposer timely and approved the launching of the "Protection Project of China Agri-cultural

---

2　Cao Xingsui. *A Discussion on the Definitions of Related Concepts of Agri-cultural Heritage*. Research on Heritage and Preservation [J]. 2016,1(1)90-94 .
3　Cao Xingsui. *A Discussion on the Definitions of Related Concepts of Agri-cultural Heritage*. Research on Heritage and Preservation [J]. 2016,1(1)90-94.
4　Li Genpan. *Development of the Agricultural History Subject and Evolution of a Concept "Agri-cultural Heritage"*. Agricultural History of China [J]. 2011, (3): 121-128.
5　Wang Siming, Lu Yong. *China's Agricultural Heritage Research: Progress and Change*. Agricultural History of China[J].2010,(1)3-11 .

Heritage" at the China Agricultural Museum. From April 2004 to June 2006, the "Agri-cultural Heritage Protection Pilot Program in Congjiang and Weining Counties of Guizhou Province" was implemented.[6] The term of "agri-cultural heritage" probably first appeared in academic journal when professor Min Qingwen published his article entitled *GIAHS: A New Kind of World Heritage* in Resources Science in 2006, which focuses on introducing the concept and project promotion of "Global Important Agricultural Heritage Systems (GIAHS)" proposed by FAO.

Agricultural history, agricultural heritage and agricultural heritage systems are intrinsically linked, but there are significant differences. Up to now, the national first-level society — China Agricultural History Association and academic journals like Agricultural History of China are still promoting the disciplinary development of agricultural history. In the field of agricultural heritage, the Chinese people have inherited traditions of the big and ancient agricultural country, and maintained the leading position in the world trend.[7] The research and practice focusing on the exploration and protection of partial FAO's GIAHS and China-NIAHS of the Ministry of Agriculture and Rural Affairs from "living" agricultural heritage provide new opportunities and impetus for the disciplinary development of agricultural history.[8]

Agri-cultural heritage is a special kind of cultural relics or cultural heritage. Cultural relics are artifacts and sites with historical, artistic and scientific values left over by human beings in social activities.[9] It is obviously that cultural relics refer to specific material remains and have two basic characteristics. First, they must be created by human beings or related to human activities. Second, they must have become the past of history and cannot be recreated.

In the international community, some international conventions on the protection of cultural relics adopted by the United Nations Educational, Scientific and Cultural Organization (hereinafter referred to as UNESCO) conferences generally refer to cultural relics as "Cultural Property" or "Cultural Heritage". The contents of the two terminologies are not the same. From the specific content listed in its conventions, the former refers to movable cultural relics, while the latter refers to immovable cultural relics.

The evolution of the Chinese Academy of Cultural Heritage can reflect some changes in the objects and ideas of protection. In 1935, the government of the Republic of China decided to set up the Old Capital Cultural Relics Arrangement Committee and its executive body — the Peiping Cultural Relics Arrangement and Implementation Office. After the founding of the People's Republic of China in 1949, the Peiping Cultural Relics Arrangement Committee (the former Old Capital Cultural Relics Arrangement Committee) and its engineering department were formally renamed the Beijing Cultural Relics Arrangement Committee. After several changes, it became the Institute of Science and Technology for Cultural Relics Protection in June 1973, and the Institute of Chinese Cultural Relics in August 1990, and was renamed the Chinese Academy of Cultural Heritage in August 2007.

It is not long before traditional cultural relics are extended to living cultural heritages for protection. In 2006, Juguan Ancient Winter Jujube Orchard and Xinjiang Karez were approved by the State Council as the sixth batch of National Cultural Relic Protection Units. The former is China's biggest existing ancient winter jujube orchard with thousands of years' history located at Juguan Village along the riverside of Niangniang River of Qijiawu County, Huanghua City of Hebei Province. The latter is also known as the three major projects of ancient China along with the Great Wall and the Beijing-Hangzhou Grand Canal. In particular, the selection of Juguan Ancient Winter Jujube Orchard opens up a new category of cultural relics protection in China and became the earliest agricultural

---

6   Cao Xingsui. *A Discussion on the Definitions of Related Concepts of Agri-cultural Heritage*. Research on Heritage and Preservation [J]. 2016,1(1)90-94.
7   Cao Xingsui. *A Discussion on the Definitions of Related Concepts of Agri-cultural Heritage*. Research on Heritage and Preservation [J]. 2016,1(1)90-94.
8   Li Genpan. *Development of the Agricultural History Subject and Evolution of a Concept "Agri-cultural Heritage"*. Agricultural History of China [J]. 2011, (3): 121-128.
9   Li Xiaodong. *Science of Cultural Relics*. Beijing: Academy Press, 2015.

and botanical National Cultural Relic Protection Units. In 2013, the Honghe Hani Rice Terraces and the Ancient Tea Plantations of Jingmai Mountain in Pu'er of Yunnan Province were selected as the seventh batch of National Cultural Relic Protection Units. In addition, Cuizhuang Ancient Winter Jujube Orchard in Binhai New District, Xinhua Ziquejie Terraced Fields, Congjiang Jiabang Terraces and Xinghua Duotian were listed as the Provincial Cultural Relic Protection Units in Tianjin, Hunan, Guizhou and Jiangsu provinces respectively. It is interesting to note that all of the above mentioned items belong to GIAHS or China-NIAHS, except the Juguan Ancient Winter Jujube Orchard.

Because the word "culture" contained in the Chinese terminology of "agri-cultural heritage", many people simply attribute agricultural heritage systems to cultural heritage. In 2004, Professor Cao Xingsui, as a member of the CPPCC National Committee, put forward the *Proposal on Incorporating of Chinese Agri-cultural Heritage into the Protection Project of Chinese Ethnic and Folk Culture*, which was formally replied by the Ministry of Culture. In 2014, the CPPCC National Committee conducted a survey on "Protecting and Utilizing China Agri-cultural Heritage" organized by its Committee of Cultural and Historical Data. In 2016, Academician Li Wenhua took the lead to propose the *Recommendations on Strengthening the Research and Protection of Agri-cultural Heritage in China*, which was approved by the national leaders in charge of cultural departments. On December 23, 2017, entrusted by the State Council, Minister of Culture Luo Shugang made a report on the work of the State Council on cultural heritage to the 31st meeting of the Standing Committee of the Twelfth National People's Congress. When talking about the establishment of a multi-level list of cultural relics and intangible cultural heritage protection, he pointed out that the identification of China-NIAHS has been carried out, and the Ministry of Agriculture has identified 91 China-NIAHS sites.

There are some initial controversies about the Chinese name for agri-cultural heritage. One important aspect of the controversy is that the relevant program name of FAO only refers to "agriculture" but "culture", about which professor Min Qingwen has an explanation.[10] As for the absence of the word "culture" in English, FAO official Mr. Parviz Koohafkan, the former coordinator of this project believes that "agriculture" is a kind of "culture" in itself. Professor Li Genpan also believes that agriculture is also a kind of culture, so it is reasonable to call "agri-cultural heritage".[11] Professor Cao Xingsui also expresses similar opinions. For the time being, he specifically refers to "agricultural heritage systems" as "agri-cultural heritage" rather than the agricultural heritage in general sense or the cultural heritage of agriculture in a narrow sense.[12]

Of course, there is another kind of intangible cultural heritage in the cultural heritage category. Among intangible cultural heritages, the most typical agricultural item is the "24 Solar Terms". In 2006, the "24 Solar Terms" was selected in the first batch of representative list of Intangible Cultural Heritage at the National Level. On November 30, 2016, it was officially selected in the list of World Intangible Cultural Heritage, becoming the 39th World Intangible Cultural Heritage of China.

Agri-cultural heritage is a special type of landscape or cultural landscape. "Landscape" was originally a geographic term. Generally speaking, it refers to the scene presented in specific area, namely visual effect. This visual effect reflects the compositional complex of the land, as well as the space and materials above the land, which is the mark on the ground made by complicated natural processes and human activities. In the *volume of Geography of the Encyclopedia of China* summarizes several understandings of landscape in geography. First, it refers to the comprehensive characteristics of a region, including natural, economic and cultural aspects. Second, it refers to general natural complexes. Third, it refers to regional units that is equivalent to the smallest first-level natu-

---

10 Min Qingwen. *Explanation of GIAHS and Other Related Questions*. Ancient and Modern Agriculture [J]. 2007,(3):116-120.
11 Li Genpan. *Development of the Agricultural History Subject and Evolution of a Concept "Agri-cultural Heritage"*. Agricultural History of China [J]. 2011, (3): 121-128.
12 Cao Xingsui. *A Discussion on the Definitions of Related Concepts of Agri-cultural Heritage*. Research on Heritage and Preservation [J]. 2016,1(1):90-94.

ral areas in the comprehensive natural regionalization hierarchy system. Forth, it refers to any regional units. [13] Landscape in ecology refers to a region of spatial heterogeneity composed of interacting patches or ecosystems, which recurs in a similar form. It is a natural complex with taxonomic implications.

The word "cultural landscape" has been widely used since the 1920s, which is the product of human activities on the land surface. It is a complex of cultural phenomena consisting of natural scenery, fields, buildings, villages, factories, mines, cities, means of transportation and roads, as well as people and clothing, etc. It reflects the characteristics of the cultural system and the geographical characteristics of a region. The American geographer Carl Ortwin Sauer advocated in his book *The Morphology of Landscape* published in 1925 that geographical features and cultural geography should be studied by observing ground scenery and cultural landscape respectively.

More attention has been paid to cultural landscape because UNESCO expanded its concept and connotation of cultural heritage; in 1992, "cultural landscape" was added as a new type (some people believe that cultural landscape still belongs to the category of cultural heritage). Among the three types of defined cultural landscapes, agriculture is closely related to the "landscape of organic evolution", which originated from the initial social, economic, administrative and religious needs, and developed to the present form through the connection or adaptation with the surrounding natural environment. In particular, one of the sub-categories — "sustainable landscape" refers to maintaining a positive social role in the society associated with the local traditional way of life, and its own evolution process is still in progress while showing the historical evidence of its evolution and development.

Although there is no "agricultural heritage" in the UNESCO World Cultural Heritage, many agricultural heritages are included in the scope of "cultural landscape". For example, the cultural landscape of Bali in Indonesia with the Subak irrigation system as its core, the Loire Valley in France with a large number of farming culture elements, the Mediterranean Agro-pastoral Cultural Landscape located in the Causses and the Cévennes, the Burgundy and Champagne vineyards with the vineyard landscape, the agricultural landscape in southern Orland, Sweden, the earliest coffee plantations in southeastern Cuba, etc.

It is interesting to note that several sites have been identified by FAO and UNESCO separately. Representatively, Rice Terraces in the Cordillera Mountains of the Philippines were identified as World Cultural Heritage in 1995, and the Ifugao Rice Terraces were selected as GIAHS in 2005. China's Honghe Hani Rice Terraces of Yunnan Province were recognized as GIAHS in 2010 and identified as World Cultural Heritage in 2013. Cultural Sites of Al Ain (and Oases Areas) in the United Arab Emirates was identified as World Cultural Heritage in 2011, and Al Ain and Liwa Historical Date Palm Oasis was identified as GIAHS in 2015. The Qanat Irrigated Agricultural Heritage System of Kasan, Iran was identified as GIAHS in 2014, and the Persian Karez was selected as the World Cultural Heritage in 2016.

2. Agri-cultural Heritage Needs to Establish New Paradigm of Conservation

Although after more than a decade of efforts, great progresses have been made in the excavation and conservation of agri-cultural heritage and good ecological, social and economic benefits have been achieved, compared with other types of natural and cultural heritages, agri-cultural heritage is still unknown popularly. Compared with the well-known natural heritages such as Shennongjia and Mount Fanjing, the cultural heritages such as the Great Wall and the Forbidden City, the mixed heritages such as Mount Tai and Mount Huang, the intangible cultural heritages such as Kunqu Opera and Kung Fu (martial arts), the agri-cultural heritage is still

---

13 Editorial Department of Encyclopedia of China Publishing House. *The volume of Geography of the Encyclopedia of China*. Beijing: Encyclopedia of China Publishing House, 1990.

"hidden out of sight". And the excavation and conservation of agricultural heritage systems still has a long way to go.

From the previous concepts of agri-cultural heritage and the identified typical items, it is not difficult to find that agri-cultural heritage has its own characteristics which means we cannot continue to use the protection ideas of cultural relics or cultural heritage in general sense, but to establish new paradigm of conservation.

The main characteristics of agri-cultural heritage can be summarized as follows. First, they are "living". Agricultural heritage systems are agricultural production systems with strong production and ecological functions, which are participated by people and still in use today. The direct production products and indirect ecological and cultural services of these systems are still the important basis for farmers' livelihood guarantee and harmonious rural development. Second, they are "dynamic". With the development of social economy and technology and to meet people's ever-increasing needs of survival and development, traditional agricultural production systems are constantly changing, but this change is the structural and functional adjustment based systematical stability. Third, they have "adaptability". With the changes of natural conditions in different regions or different historical periods, the traditional agricultural production systems are constantly changing in adaptability, which is the co-evolution based on systematical stability. They can fully reflect the wisdom of harmonious survival between man and nature. Forth, they have "complexity". Agricultural heritage systems include not only the traditional agricultural knowledge and technology in general sense, but also those traditional agricultural landscapes with long history and reasonable structure, as well as unique agrobiological resources and rich biodiversity. They embody the complex characteristics of natural heritage, cultural heritage, cultural landscape heritage and intangible cultural heritage. Fifth, they are "strategic". Agri-cultural heritage systems are of great strategic significance to cope with the impact of globalization and global change, to protect biodiversity, to ensure ecological security and food security, to alleviate poverty effectively, to promote sustainable agricultural development and the construction of rural ecological civilization. Sixth, they are "multi-functional". Agricultural heritage systems have diversified material production and enormous ecological and cultural value, which fully reflects the functions of food security, raw material supply, employment and income increase, ecological conservation, sightseeing and leisure, cultural heritage and scientific research, etc. Seventh, they are "sustainable". After thousands of years of inheritance, agricultural heritage systems are one kind of typical "social-economic-natural" complex system, which consists of structural-reasonable ecosystem, diversified-output economic system as well as harmonious and stable social system. They are the model for harmonious and sustainable development of human and land. Eighth, they are "endangered". Mainly due to policies, technological reasons and comparative benefits, etc., traditional agricultural systems are facing various threats in the reduction and loss of agrobiodiversity, the disappearance of traditional agricultural technology and knowledge system and the destruction of the structure and function of agricultural ecosystem.[14]

The following are the basic conservation principles for agri-cultural heritage systems. Agricultural heritage systems are systematic heritages based on the function of agricultural production. The basic conservation principle is "conservation in excavation, and inheritance in utilization". Specifically, it concludes "conservation as priority, appropriate utilization, overall conservation, coordinated development, dynamic conservation, adaptive management, living conservation, functional expansion, in-situ conservation, demonstration and promotion, multi-participation and benefits sharing". The core principle is "whole, dynamic, living".

With the development of social economy and the improvement of people's living standards, agricultural production is also progressing. The structure, elements and functions of agricultural production systems are

---

14 Min Qingwen. *The Concept, Characteristics, Conservation and Development of Agri-cultural Heritage*. Famers' Daily. 8 February 2013, the 4th page.

changing accordingly. It is difficult for agricultural heritage systems to truly testify their "original features". Moreover, the agricultural heritage systems are mostly located in economically backward, ecologically fragile and culturally abundant areas, so there are multiple tasks of economic development, ecological conservation and cultural inheritance. Overemphasizing the conservation of "original features" while ignoring regional development, it is difficult to mobilize the enthusiasm of local residents for heritage conservation, and to achieve the purpose of conservation. It is necessary to explore new ideas of dynamic conservation and adaptive management.

To change or not to change is important aspect of agricultural heritage systems conservation. "Changeability" is absolute, and "unchangeability" is relative. The key is what can be changed (noncritical elements of agricultural heritage systems can be changed), what cannot be changed (critical elements of agricultural heritage systems cannot be changed), and how to grasp the "extent" of the changes. The so-called dynamic conservation means that we should not overemphasize the "original feature" or unchanged "frozen preservation", but should adjust appropriately in accordance with the actual situation to realize the "living inheritance" and "sustainable development" of agricultural heritage systems. However, the basic structure and function of the systems, as well as their important species resources, agricultural landscape, water and land resources management technology, etc. should not be changed, and its relevant national culture and traditional knowledge should not be changed to a large extent.[15] Without conservation, the foundation of development will be lost and the development cannot be talked about; without development, the motive force of conservation may be lost and the purpose of conservation will be difficult to achieve.

The following discusses the key conservation mechanisms of agricultural heritage systems. First, it is necessary to establish a "policy incentive mechanism" for agricultural heritage systems conservation. Agricultural heritage systems contain biological, cultural and technological "genes" of great value for present and future agricultural development: 1) Many important animal and plant genetic resources and related biodiversity in traditional agricultural production systems play an important role in maintaining ecosystem stability and services; 2) The rich and colorful songs and dances created in the process of agricultural production, such as Dong's Grand Song, Hani's Four Seasons Production Tune and Qingtian Fish Lantern Dance, as well as the material cultural heritages and intangible cultural heritages such as folk customs, diet and architecture, are of great significance to the inheritance of farming culture and the harmony of rural society; 3) Traditional agricultural production technologies, such as rice-fish system, mulberry dyke & fish pond, agroforestry, terraces tillage and intercropping are of great significance for the development of modern ecological agriculture. Therefore, it is necessary to study and implement the ecological and cultural compensation for agricultural heritage systems conservation, that is, to make ecological compensation for the conservation of agricultural biodiversity and agroecological landscape in the light of the ideas and practices for natural ecology conservation, and to make cultural compensation for the conservation of agricultural technology and cultural diversity in the light of the ideas and practices for the conservation of cultural relics, intangible cultural heritage and traditional villages.

Second, it is necessary to establish an "industry promotion mechanism" for agricultural heritage systems conservation. In addition to the direct production function, agricultural heritage systems also have important ecological and cultural functions, which lays a resource foundation for expanding agricultural functions, promoting agricultural quality and efficiency, increasing farmers' employment and income, and increasing rural harmony and stability. Agricultural heritage systems have innate advantages in developing "the sixth industry". The unique agricultural species and biological resources, relatively abundant labor resources, traditional cultural customs and beautiful rural landscapes have become the

---

15 Min Qingwen. *The Dynamic Conservation of "Five in One" -- Who Will Conserve and How to Conserve them?* China Profiles [J]. 2016,(1):42-44.

advantages of developing labor-intensive characteristic agriculture and agricultural products processing industry, handicraft production industry, ecological and cultural tourism, biological resources industry, cultural and creative industries, etc. On the premise of adhering to agricultural production as the basis, we should actively develop multi-functional agriculture by developing ecological agricultural products with cultural connotations, agricultural products processing industry, food processing industry, biological resources industry, cultural and creative industries, rural tourism, etc. We should gradually establish a new agricultural industry model of agricultural function expansion and integration of "agricultural production, agricultural product processing industry and agricultural product market service industry", so as to realize the transformation of farmers' role from "agricultural producers" to "multiple operators", and the transformation from agricultural activities, rural landscape, traditional folklore, ecological environment to ecological and cultural tourism resources, and from self-sufficient agricultural products to characteristic agricultural products with higher added value, high-end consumer goods and tourist souvenirs.

Third, it is necessary to establish a "multi-participation mechanism" for agricultural heritage systems conservation. Agricultural heritage systems are traditional agricultural production systems created, inherited and developed by the ancestors. Their owner should be "farmers" who are still engaged in agricultural production. They should be the main protectors of agricultural heritage systems and the main beneficiaries from them. However, it must be noted that the reason for agricultural heritage systems conservation is precisely because they are facing threats, uncompetitive and in an "endangered" state. If the conservation totally depends on farmers, it is not only difficult to achieve the purpose of conservation, but also unfair to place the burden of the "heritage" conservation, which belongs to the common sharing of all mankind, on the vulnerable groups.

Therefore, a "five-in-one" multi-party participation mechanism should be established, which includes government promotion, science and technology driving, enterprise leading, community initiative and social linkage. The government should play a leading role in formulating relevant supportive policies, implementing standardized management, organizing the compilation of plans and their implementation, and taking charge of fund raising, so as to incorporate the conservation and utilization of agricultural heritage systems into the overall layout of local development. The supporting role of science and technology should be exploited to organize experts in the fields of agroecology, agricultural history, agri-culture, agricultural economy and rural development to explore and evaluate the value of agricultural heritage systems, analyze the sustainable development mechanism of agricultural heritage systems, assist in formulating scientific and operational conservation and development plans, and carry out theoretical upgrading of traditional knowledge and experience. The initiative of rural communities to conserve agricultural heritage systems should be fully mobilized to enhance their self-confidence so that they can truly become the main beneficiaries of the results of agricultural heritage systems conservation. The special role of enterprises in the conservation and development of agricultural heritage systems should be emphasized to effectively improve the level of product development, market expansion, capital investment and industrial management. Public participation awareness should be improved to create good social atmosphere and explore community support for agri-culture, adoption system, volunteer system and other measures.[16]

3. Image Recording Can Effectively Promote Agricultural Heritage Systems Conservation

Image recording plays an important role in the protection of natural and cultural heritage. In recent years, the protection of natural and cultural heritage by means of image technology has attracted much attention. For example, in order to promote the protection and inheritance

---

16 Min Qingwen. *The Dynamic Conservation of "Five in One" -- Who Will Conserve and How to Conserve them?* China Profiles [J]. 2016,(1):42-44; Min Qingwen. *Key Mechanisms of Agricultural Heritage Systems*. Guangming Daily. 19 August 2016, Page 10.

of intangible cultural heritage, the General Office of the State Council in its *Opinions on Strengthening the Protection of China's Intangible Cultural Heritage* (issued on March 26, 2005) stressed that "the intangible cultural heritage should be recorded authentically, systematically and comprehensively by means of words, sound recordings, video recordings and digital multimedia to establish archives and databases." Many scholars or students have studied and discussed the significance of images for the protection of cultural heritage. Cui Ying believes that in the context of "intangible cultural heritage" protection, because of its powerful "audio-visual integration" function, image has obviously become the communicator for its indirect protection function of intangible cultural heritage by means of communication. This mode of communication can be regarded as "ritual" and image isomorphism, i.e. showing "ritual" representation in a specific cultural space, while drawing lessons from the expression paradigm of image deep description and encoding to form a close combination, so that the image can be placed into different "fields" and merged with "intangible cultural heritage" in a freer role to disseminate refined and transformed cultural elements or representative symbols of regional culture.[17] Lin Lishun's exposition of the advantages of using images to protect intangible cultural heritage is of great significance for reference. First, vivid and intuitive images are more convenient for people to understand the core content of intangible cultural heritage, and can better protect the original ecology of intangible cultural heritage. Second, images can help to transform intangible cultural heritage into substantial form for long-term preservation. Third, images can transform most of the cultural items of intangible cultural heritage into the image expression pattern, covering a wider range and can record the social and cultural context of these intangible cultural heritage in a more complete and in-depth way within the all-round and three-dimensional intuitive exposition system. Fourth, the communication channels of image expression are wider, which can be disseminated in the form of feature films and documentaries, and can also be used as long-term preservation of image archives. Fifth, the method of image expression can provide the image basis for reconstruction and re-development of local cultural traditions, and provide the image basis for the living protection of intangible cultural heritage. From the perspective of identity, the protection and wide dissemination of intangible cultural heritage by images can not only strengthen public collective identity, but also affect public cultural and social identity. It can lay a solid foundation for our future generations to have a good collective memory of the intangible cultural heritage, which is undoubtedly a benign adaptation to the rapid changes of the times. On one hand, they can increase their sense of pride. On the other hand, they will be more willing to protect intangible cultural heritage and consciously inherit and re-develop the traditional Chinese culture. The cultural identity and ethnic identity of Chinese people can be strengthened through the images of film and television media's symbols and connotation systems.[18]

Although static photos do not have the function of dynamic images, they can also play an important role in archival records, widespread dissemination, reader acceptance, increasing cultural identity and cultural self-confidence in heritage protection. *The World Heritage in China* edited and published by China Photographic Publishing House in 2014 is a good example. It shows the unique charm of China's world heritages from various angles in an in-depth way with both pictures and texts, which can be best testified by several experts' comments. Liu Jiaqi, an academician of the Chinese Academy of Sciences, believes that this book is a classic of culture and art, which gathers the essence of China's nature and culture. Ge Jianxiong, a historic geographer, told the reader, "If you can't go for witnessing at present, you can travel around with your mind by this book. If you are ready to go, you may hold this book as a guide." Zhou Meisheng, an image expert of the world heritage site, believes that the book is "the

---

17 Cui Ying. *Role and Significance of Video Imaging in the Protection of Intangible Cultural Heritages*. Journal of Yunnan Nationalities University (Philosophy and Social Sciences). 2018, 35(6):38-42.
18 Lin Lishun. *The Way of Imagery Protection of Intangible Cultural Heritage*. Southeast Communication [J]. 2014,(10):54-56.

image proof of the world heritage sites, the historical map of Chinese civilization, and the visual symphony between man and nature".[19]

Image recording plays a unique role to conserve agricultural heritage systems. Because agricultural heritage systems are living traditional agricultural systems with productive functions, systematic conservation should be the first choice, that is, through the identification of important agri-cultural heritage sites to define the core areas of conservation, determine the key elements of conservation, formulate effective measures for conservation, transform single agricultural production into multi-functional agricultural development, in order to allow farmers to gain benefits through participation in conservation, allow agriculture continuously develop by improving efficiency, and allow heritages to be inherited and flourished by sustainable development. However, because agricultural heritage systems are dynamic evolution systems, the change of some elements is unavoidable. It is very important to preserve the elements or components that may change by using (ecological) museums or (species) resource banks to record them by means of images and information, as well as establishing heritage bases.

Therefore, as far as agricultural heritage systems are concerned, image recording and conservation is one kind of method. Although it is neither the best nor the only way, it can play its unique role. First, it can play the role of "preservation" by image archiving the elements and landscapes which may be difficult to retain in the living pattern, so as to give full play to their literature and academic value. Second, it can overcome the seasonal drawbacks of agricultural production and agricultural landscape, so that people can understand the scenes of different periods and seasons at the same time. Third, it can play a role in scientific transmission, making more people vividly understand agricultural heritage systems, in particular for those who have no chance to visit the sites in person to feel the charm of agricultural heritage systems to a certain extent. Fourth, it can play the role of popular science education, so that more people can perceive the rich connotation of agricultural heritage systems through photos.

Publishing Image Recording of Agricultural Heritage Systems in Chinese and English versions will help to promote China's excellent farming culture to the world. China is one of the most important origins of agriculture in the world. It has a history of farming culture for thousands of years. Agricultural civilization is not only the foundation of ancient Chinese civilization, but also has a far-reaching impact on the development of global agricultural civilization.

At present, the exploration and conservation of agricultural heritage systems has achieved wider international consensus. In view of the problems brought about and faced by the development of modern agriculture, such as large consumption of resources, ecosystems degradation, biodiversity loss, serious environmental pollution, loss of traditional technology and resistance to the inheritance of traditional culture, the Food and Agriculture Organization of the United Nations (FAO) with the support of the Global Environment Facility (GEF) and in conjunction with relevant international organizations and countries jointly launched the initiative of Global Important Agricultural Heritage Systems (GIAHS), aiming at the establishment of conservation systems for global important agricultural heritage systems and its related landscape, biodiversity, knowledge and culture to ensure their recognition and conservation worldwide, making them the foundation of sustainable management. The initiative will strive to promote a better regional and global understanding of the traditional knowledge and management experience of local farmers and ethnic minorities on nature and environment, and to use these knowledge and experiences to meet the challenges of contemporary development, in particular to promote the revitalization of sustainable agriculture and the realization of rural development goals. Since 2005, five different types of traditional agricultural systems have been selected as the first batch of conservation pilot sites in six countries. After more than a decade of efforts, 57 sites in 21 countries have been included in the list of GIAHS.

---

19 Zhao Yingxin Editor in Chief. *World Heritage in China*. Beijing: China Photographic Publishing House. 2014

China is the earliest responder, most active participant, strongest supporter, most successful practitioner, most important promoter and main contributor of FAO's GIAHS initiative. A few examples can be cited to illustrate this as follows. First, in 2005, the Qingtian Rice-Fish Culture System of Zhejiang Province was successfully recommended as the first batch of pilot sites and formally certificated. Up to now, 15 GIAHS sites have been certificated by FAO, ranking first among all countries. Second, in 2012, China took the lead in excavating Nationally Agricultural Heritage Systems, and the Ministry of Agriculture and Rural Affairs has identified four batches of 91 China-NIAHS sites. Third, in 2013, China initiated the establishment of the East Asia Research Association for Agricultural Heritage Systems (ERAHS) and took the lead in organizing regional academic seminars. Forth, in 2014 and 2016, China successfully promoted the incorporation of GIAHS into the *Asia-Pacific Economic Cooperation (APEC) Declaration of the Ministerial Conference on Food Security* and the *Declaration of the Conference of Agricultural Ministers of the Group of Twenty (G20)*. Fifth, China has continuously organized "High Level Training Courses on GIAHS" since 2014, which has trained more than 100 trainees from over 60 countries and international organizations. Many GIAHS sites liking Qingtian, Shaoxing and Huzhou of Zhejiang Province, Longsheng of Guangxi Zhuang Autonomous Region, Xiajin of Shandong Province, Xinghua of Jiangsu Province, Fuzhou and Youxi of Fujian Province etc. have become important training and education bases. Sixth, the Agricultural Heritage Systems Branch of the China Association of Agricultural Science Societies was found in 2014, which held annual academic exchanges and experiences sharing continuously. Seventh, in 2015, China promulgated the *Measures for the Management of Important Agricultural Heritage Systems*, which made China the first country to promulgate management measures on this field. Besides, GIAHS monitoring and evaluation began in the same year. Eighth, in 2016, a survey of agri-cultural heritage was carried out nationwide, and 408 potentially agricultural heritage systems were identified that year. Ninth, Academician Li Wenhua was the first chairman of the GIAHS Project Steering Committee in 2011. Tenth, professor Min Qingwen was awarded the first "Special Contribution Award for GIAHS" by FAO in 2013. Eleventh, Mr. Jin Yuepin from Qingtian County was awarded the title of "Model Farmer in Asia-Pacific Region" in 2014 for his contribution to the conservation of agricultural heritage systems. Twelfth, professor Min Qingwen was elected as the first chairman of GIAHS Scientific Advisory Group (SAG) in 2016, etc.

In a sense, China's GIAHS and NIAHS conservation are affecting the sustainable development of global agriculture and rural revitalization. Chinese experiences in exploring and conserving agricultural heritage systems has led and will continue to lead the international movement for agricultural heritage systems conservation. The publication of this book will undoubtedly help to further promote the excellent farming culture of China to the world.

Agricultural heritage systems are very important. The comments below may be the best explanation of their significance: "Farming culture is the precious wealth of Chinese agriculture, which is an important part of Chinese culture. It cannot be lost, but to be continuously developed and flourished."

Let us remember the slogan of FAO: GIAHS is not for the past, but for the future.

Min Qingwen
April 2, 2019

(The author is a member of the CPPCC National Committee, Co-Chairman of the FAO's GIAHS Scientific Advisory Group, Vice-Chairman and Secretary-General of the GIAHS/China-NIAHS Expert Committee of the Ministry of Agriculture and Rural Affairs, Vice-Chairman of the Ecological Society of China, Vice-Chairman of the China Agricultural History Association, Vice-Chairman and Secretary-General of the Agricultural Heritage Systems Branch of the China Association of Agricultural Science Societies, Professor of the Institute of Geographic Sciences and Natural Resources Research of the Chinese Academy of Sciences, Vice Director of Center for Natural and Cultural Heritage)

华北地区·North China

天津市

皇家枣园的代表——天津滨海崔庄古冬枣园

河北省

传统漏斗架葡萄栽培体系——河北宣化城市传统葡萄园

北方林粮间作典型模式——河北宽城传统板栗栽培系统

北方梯田与石头文化的集成——河北涉县旱作梯田系统

内蒙古自治区

世界旱作农业源头——内蒙古敖汉旱作农业系统

蒙古族游牧文化的缩影——内蒙古阿鲁科尔沁草原游牧系统

皇家枣园的代表
天津滨海崔庄古冬枣园

中国人爱枣,这种喜爱贯穿了枣的一生,从"枣芽初长麦初肥",到"簌簌衣巾落枣花"。枣树结果了,更是高兴,"行过大山过小山,房上地下红一片""河上秋林八月天,红珠颗颗压枝园",足以窥见一斑。中国是枣的起源地,关于枣的记述,最早可以追溯到《诗经》,"八月剥枣,十月获稻"。成书于汉代的《尔雅》,在其《释木》一篇中,第一次解释和记录了枣的品种。根据其中的记述,在我国周代,枣的品种就多达11种。

枣树的适应性极强。众多资料显示,枣树对环境要求不高,它耐旱、耐涝、耐贫瘠、耐盐碱。从我国东部的山东沾化大枣,到西部的新疆哈密大枣,枣树这种"好养活"的习性,使其产区遍布大半个中国。

现天津滨海新区大港太平镇崔庄村,与紫禁城的直线距离只有160千米。这里占地3000亩的冬枣园被称为"皇家枣园",是我国成片规模最大及保留最完整的古冬枣林,被认为是"皇家枣园"的代表。据史料记载,在600多年前的明代,人们就开始在这片区域种植冬枣树。相传,明孝宗朱祐樘及其皇后曾经在这里采摘、品尝过冬枣,"皇家枣园"的名号便由此得来。事实如何虽已无从考究,但是,冬枣园里的168棵600年以上的枣树、3232棵400年以上的枣树却是真实存在。现如今,"皇家"没了,枣园还在。

坑洼不平的冬枣树干写满历史,是现今研究枣树起源、演化和枣文化的"活文物"。每年果期,新鲜的、累累的果实,让人丝毫猜不出这些枣树的年纪,而其果实皮薄核小肉厚、肉质酥脆甘甜的特点,也让它丝毫不愁没有市场。

地理位置:北纬38°33′—38°57′、东经117°08′—117°34′,地处天津市滨海新区大港太平镇崔庄村。

气候特点:温带大陆性季风气候。

认定时间:2014年被农业部列为第二批中国重要农业文化遗产。

Geographical location: 38°33′-38°57′N, 117°08′-117°34′E; located in Cuizhuang Village, Taiping Township, Dagang District, Binhai New District, Tianjin.

Climate type: temperate continental monsoon climate.

Time of identification: In 2014, it was listed as one of the second batch of the China-NIAHS by the Ministry of Agriculture.

A Representative of Royal Jujube Orchard
Cuizhuang Ancient Winter Jujube Orchard

Chinese people's affection runs through every stage of the life of jujubes, from "the buds breaking above the soil", like one sentence written in an ancient poetry, to "jujube flower withering and falling on clothes and scarves". Men of letters got even more excited when they saw the jujubes fruiting: "Crossing the mountains and hills, harvesting the jujubes here and there"; "Forests of autumn touch the skies, beads in red drizzle your eyes." China is the origin of jujubes. The description of jujube can be traced back to *The Book of Songs*, in which there is a line reading "Peel jujubes in August and harvest rice in October." One passage named *Interpretation of the wood* (mainly explains something about plants) is part of the book *Erya* (written in the Han Dynasty; one of the classics of Confucianism), in which the varieties of jujubes were explained for the first time. According to the description of it, there were as many as 11 varieties of jujubes in the Zhou Dynasty.

Jujube trees are highly adaptable. Numerous researches show that the trees are not demanding on the environment. Instead, they are drought-tolerant, sturdy, resistant to barrenness, salt and alkali. From the Zhanhua jujubes in the east of China to the Kumul jujubes in the west, the "easy-to-cultivate" capacity of jujube trees make their production area spread over half of China.

The Cuizhuang Village in Taiping Township, Dagang District, Binhai New District, Tianjin is only 160 kilometers away from the Forbidden City. The jujube orchard here, covering an area of 3,000 mu (1mu=0.0667 hectares), is well known as the "Royal Jujube Orchard". It is the largest and most intact ancient winter jujube forest in China, and is considered to be the representative of the "Royal Jujube Orchard". According to historical records, in the Ming Dynasty more than 600 years ago, people began planting winter jujube trees in this area. There is a folklore going that in the Ming dynasty, the emperor Zhu Youtang (his posthumous tile is Xiaozong) and his empress once picked and tasted the winter jujubes here, and then the name of the "Royal Jujube Orchard" came from that. Though the fact about the story cannot be investigated any more, the 168 trees in the orchard with more than 600 years and 3,232 trees with more than 400 years are real. Nowadays, the royal family is gone, while the jujube orchard is still there.

The rutted ancient winter jujube trunks full of history are the "living cultural relics" for helping researchers to study the origin, evolution and culture of jujubes. Every year when it comes to the jujube harvest, people will wonder the age of these jujube trees because the fruits are so fresh and crowded around the branches that it seems the oldness of these trees don't deserve them. Thin the skin of its fruits is, small the nutletes are and thick and crispy and sweet sarcocarp they have, the jujubes do not need to worry their market destiny at all.

滨海崔庄古冬枣园位于天津市滨海新区大港太平镇崔庄村，毗邻荣乌高速公路，面积约 3000 亩，其中古冬枣核心区面积 238 亩，是我国成片规模最大及保留最完整的古冬枣林。（焦永普 摄）

The ancient winter jujube orchard in the picture is located in Cuizhuang Village, Taiping Township, Dagang District, Binhai New District, Tianjin City. It is adjacent to Rongwu Expressway (from Rongcheng city in Shandong Province to Wuhai City in Inner Mongolia Autonomous Region). It covers an area of about 3,000 mu, including the core area of the ancient jujube trees of 238 mu, which makes it the largest and most intact ancient winter jujube forest in China. Photographed by Jiao Yongpu

天津滨海崔庄古冬枣园是以明清"皇家枣园"为特色的古冬枣栽培系统。图为今天复建的冬枣园外景和枣园里枣树下休憩的老者和儿童摘枣雕塑。（焦永普 摄）

The ancient winter jujube orchard in Cuizhuang village, Binghai New District employs the cultivation system of ancient winter jujube trees which featuring the "Royal Jujube Orchard" in the Ming and Qing dynasties. The pictures here show the exterior of the ancient winter jujube orchard built today and two sculptures of an elderly man resting under a jujube tree and a boy picking the jujubes. Photographed by Jiao Yongpu

古老的冬枣树曾因"大炼钢铁"而被伐薪烧炭。值得庆幸的是,在崔庄有识之士的庇护下,少量成片古冬枣树得以保留,终将古冬枣树这一珍贵资源留存至今。图为树龄达600年以上的"嫡祖"树,以及古树虬枝。(上图 刘岳坤 摄;下图 焦永普 摄)

The ancient winter jujube trees were once burned down to the ground because of the "universal steelmaking" event during the Great Leap Forward movement. Fortunately, under the protection of the persons with breadth of vision in Cuizhuang Village, a small amount of ancient winter jujube trees were preserved and the precious resource of the jujubes could be survived to this day. The pictures show the "ancestor" of jujube trees and its interlaced canes, which is more than 600 years old now. The picture above photographed by Liu Yuekun, and below by Jiao Yongpu

近年来，大港地区不断发掘冬枣文化，推出了崔庄"冬枣文化节""枣花姑娘评选"等活动，吸引了大量中外游客。（刘岳坤 摄）

In recent years, Dagang District has continuously explored the culture of winter jujubes, and launched activities such as the "Winter Jujube Culture Festival" and "Jujube Girl Appraisal" in Cuizhuang Village, which attract a large number of domestic and foreign tourists. Photographed by Liu Yuekun

随着大港地区冬枣种植规模的扩大，崔庄冬枣通过了"国家地理标志产品""无公害农产品"认证。图为夏秋之交，带着火红太阳色的大枣挂满枝头，枣农采摘收获的场景。（刘岳坤 摄）

With the expansion of planting scale in Dagang Distric, winter jujubes here pass the certification of "China Protected Geographical Indication Products" and "Pollution-free Agricultural Products". At the end of spring and the beginning of summer, the flaming red jujubes hang on the branches. The pictures show the scene of the harvesting. Photographed by Liu Yuekun

为加强对崔庄古冬枣园的保护管理，滨海新区已经编制了系列保护办法和规划，对古冬枣园进行统一管理和保护，确保滨海崔庄古冬枣园这一宝贵财富得以传承和发展。古老的冬枣依然焕发着生机与活力。（焦永普 摄）

In order to strengthen the conservation and management of the Cuizhuang ancient winter jujube orchard in Cuizhuang Village, authorities in Binhai New District has compiled a series of protection measures and plans and carry out unified management and protection to ensure the inheritance and development of the precious wealth of the ancient winter jujube orchard. The jujubes are still full of vitality. Photographed by Jiao Yongpu

传统漏斗架葡萄栽培体系
河北宣化城市传统葡萄园

宣化是历史文化名城，而在内涵丰富、积淀深厚的宣化文化中，葡萄也扮演着重要的角色。这座位于首都北京西北方向150千米左右的古城，甚至用葡萄来为自己命名——"葡萄城"。根据文献记载，从唐代起，宣化就开始栽种葡萄，距今已有超过1300年的历史。

中国最早关于"葡萄"的文字记载出现于《诗经》，里面提到"六月食郁及薁"。其中，"薁"说的就是野葡萄。这与我们今天所食用的葡萄有所不同，现在指的葡萄，是西汉时期，张骞出使西域，带回的当时已由欧洲传入西域的欧洲葡萄——没错，我们今天吃的葡萄，发源地在欧洲，这也是宣化葡萄的"祖先"。

宣化开始种植葡萄，距离葡萄传入中国已经过了800多年，这反而让宣化有了后发优势，拥有了种植经验和较为成熟的种植方法。宣化传统的漏斗架葡萄栽培方式延续至今：葡萄架被围成一个圆圈，架身向上方倾斜30°—35°，内圆外方，呈放射状，整个架式像一个漏斗，"漏斗架"也因此而得名。这种架式占地面积小，可以使光能、肥源、水源集中，还有抗风、抗寒的特点，在国内乃至世界上都是独一无二的。

不知是古人的智慧，还是一种因缘际会的巧合，漏斗架与宣化的自然地理条件惊人相似：四周群山环抱，如向上方倾斜的架身。盆地地形使这里昼夜温差大，从区域中心穿过的河流，给予了葡萄种植所需的充足水源。一个"大漏斗"，里面套着无数"小漏斗"，在漏斗套漏斗中生产出来的葡萄，皮肉呈黄绿色，口感脆而多汁，酸糖比适中，有着"刀切牛奶不流汁"的美誉。

在漫长的岁月中，传统漏斗架葡萄栽培在宣化人中代代传承，不断优化。这也是让宣化人骄傲的地方。

地理位置：北纬40°37′、东经115°03′，地处河北省张家口市宣化区。
气候特点：中温带亚干旱气候。
认定时间：2013年被联合国粮农组织列为全球重要农业文化遗产，同年被农业部列为首批中国重要农业文化遗产。

Geographical location: 40°37′ N, 115°03′ E; located in Xuanhua District, Zhangjiakou City, Hebei Province.
Climate type: subtemperate subarid climate.
Time of identification: In 2013, it was listed as one of the GIAHS by the FAO. In the same year, it was listed as one of the first batch of the China-NIAHS by the Ministry of Agriculture.

Traditional Funnel Rack Viticulture System

Urban Agricultural Heritage of Xuanhua Grape Gardens

Xuanhua, located about 150 kilometers northwest of Beijing, the capital, is a famous historic and cultural city. In the cornucopia of the rich and profound culture, grape is one of the plumpest fruits, and the ancient city even names herself The City of Grape. According to the documentary records, Xuanhua began to plant grapes since the Tang Dynasty, which enjoys a history of more than 1,300 years.

The earliest written record of "grape" in Chinese is *The Book of Songs*, that is: "Eat the Japanese bush cherry and the black nightshade herb in June." The black nightshade herb is a kind of Chinese herbal medicine, and one of its aliases is "wild grape", while different from the grapes we eat today, which were the one that were introduced from Europe to the West Regions and brought in China by Zhang Qian when he was served as an envoy abroad during the Western Han Dynasty, and also the one that "gave birth" to the grapes in Xuanhua.

It seemed that Xuanhua was fell behind since it started planting grapes more than 800 years later after Chinese knew what grapes were. Instead, thanks to this, more planting experiences and mature cultivation methods could be mastered, which helped the latercomer surpass the formers. A traditional funnel rack viticulture system continues up to now: the grape rack is formed a circle, with its body inclines upwards by 30-35 degrees. It is roundish inside while square outside, and in a radial pattern, which make it like a funnel on the whole. "Funnel rack" is also named for that. This type of rack has a small footprint, which can concentrate light, fertilizer and water and resistant to wind and cold. It is unique in China and even the whole world.

Whether it is the wisdom of the ancients or a coincidence of chance, funnel racks are strikingly similar to the natural geographic environment in Xuanhua. The city is surrounded by mountains, which is just like the rack body inclined upward. Xuanhua enjoys a large difference in temperature between day and night because of the basin terrain, and softly watches the river running though the region centre and feeding the grapes with sufficient water. Blessed by the "big funnel" with countless "small funnels" in it, the skin of those grapes is yellow-green, and our mouth surely can feel the crisp and feast on its juicy with pleasant sweetness and sourness. It is well-known that the juice in grapes is so thick that it is as solid as a milk curd and does not flow when cut it.

In the river of time, the traditional funnel rack viticulture cultivation method has been inherited down from generation to generation and optimized step by step, which exists what Xuanhua people are proud of.

坐落于首都北京西北约150千米处的宣化古城,历来就有"葡萄城"的美誉。每年中秋前后,满城葡萄飘香,串串晶莹剔透的牛奶葡萄吸引着八方来客。图为航拍宣化"葡萄城"。(杨利华 摄)

The city, seated 150 kilometers northwest of Beijing, has enjoyed a high reputation of "The City of Grape" since long. Every around the Mid-Autumn Festival, thousands of branches of the crystal "milk grapes" will send fragrance to the city and attract numerous visitors from every corner of the world. The picture is the aerial scene of "The City of Grape". Photographed by Yang Lihua

据《宣化葡萄史话》记载，宣化葡萄最早引进栽培的时间为唐代。如今，在宣化古城的观后村里，有一株近 600 岁的古葡萄藤，依然枝繁叶茂、硕果累累，见证着宣化葡萄发展的历程。（赵占南 摄）

A book named *A Brief History of Xuanhua Grapes* recorded that grapes were first introduced and cultivated there in Tang Dynasty. Today, an ancient grape vine behind Guanhou village, which is nearly 600 years old while still flourishing and fruitful, witnesses the development of Xuanhua grapes. Photographed by Zhao Zhannan

宣化传统葡萄园至今仍沿用传统的漏斗架栽培方式。漏斗架是一种古老的传统架式，因其架式像漏斗而得名。图为传统的葡萄树每年都在农人的悉心照料下焕发新的生机。（赵占南 摄）

The traditional vineyards in Xuanhua still use the traditional funnel rack viticulture method. The funnel rack has a long history, which is named after its frame like a funnel. The traditional vines are rejuvenated every year with the care of the farmers. Photographed by Zhao Zhannan

近600岁的古葡萄藤号称"天下第一老藤",人们每年除了品尝鲜美的葡萄,也习惯于在老藤上系上自己的祝福与祝愿。不仅如此,"京西第一老藤"每年都吸引国内外游客前来参观。(赵占南 摄)

Nearly 600 years old, the ancient vines are known as "the eldest vine in the world". Every year, in addition to tasting the delicious grapes, people are used to attaching their wishes to the old vines. Not only that, "the eldest vine in western Beijing" attracts domestic and foreign tourists to visit every year. Photographed by Zhao Zhannan

宣化牛奶葡萄属鲜食葡萄品种，皮肉黄绿色，质脆而多汁，酸甜比适中，素有"刀切牛奶不流汁"的美誉。（赵占南 摄）

The Xuanhua grapes, the variety of table grapes, with yellow-green skin, crisp and juicy taste and moderate acid sweet, thus enjoy a reputation of "being cut into slices without losing the juice". Photographed by Zhao Zhannan

如今葡萄及其架式已经成为宣化的代称和标志。图为矗立在宣化古城的巨型葡萄玉质雕塑和模仿漏斗葡萄架的地标建筑。（孙辉亮 摄）

Today, the grape and its funnel rack have become the alternative name and symbol of Xuanhua. The pictures show the giant grape jade sculpture standing in the ancient city of Xuanhua and landmark building that imitates the funnel grape rack. Photographed by Sun Huiliang

1988年9月,张家口市宣化区人民政府举办了第一届"中国宣化葡萄节"。节日期间,以葡萄为媒介,文化搭台、经济唱戏,邀请各界知名人士,品尝葡萄美味,感受古城文化,洽谈建设项目,带动经济发展,为葡萄产业赋予了浓郁的文化内涵。图为今天的宣化葡萄成熟、采摘及销售盛况。(赵占南 摄)

In September, 1988, the people's government of Xuanhua District, Zhangjiakou City held the first "Xuanhua Grape Festival". During the festival, thanks to the grapes who establish a stage for showing the Xuanhua culture to promote the economy, celebrities from all walks of life are invited to taste the delicious grapes, enjoy the culture of the ancient city, negotiate their projects, which have tremendously promoted the economic development and gave the grape industry a rich cultural connotation. Pictures here show the spectacular scene of ripen grapes, and numerous people who are occupied to pick and sell grapes. Photographed by Zhao Zhannan.

2018年9月23日至25日,第一届中国农民丰收节暨第三届宣化观后村葡萄采摘节在中国宣化莲花葡萄小镇举行。(孙辉亮 摄)

On September 23-25th, 2018, the first Chinese Farmers Harvest Festival and the third Xuanhua Guanhou Grape Picking Festival was held in Xuanhua Lianhua Vine Town. Photographed by Sun Hui-liang

北方林粮间作典型模式
河北宽城传统板栗栽培系统

"板栗烧肉"是江浙地区餐桌上的一道家常菜,而这道菜的主材料——板栗,到了北方,则是人们茶余饭后的一种消遣零食。无论是翻滚在街边小巷的黑铁锅里,还是被置于商场精美的零食柜里,板栗那经过不断翻炒和烘焙后散发出的香甜味道,总能在隆冬给人带来一丝安慰。

在中国,板栗已有几千年的种植历史。作为板栗的原产地,中国目前保有300多个板栗品种。现在的辽、蒙、苏、鲁、陕等中国的大部分省区,都能看到板栗树的踪影,板栗年产量居世界首位。这其中,数燕山山脉一带出产的板栗质量最为上乘。燕山深处的宽城,土壤含铁量高、阳光充足、昼夜温差大,十分适宜板栗生长。这里已有2000多年的板栗种植历史。据传,康熙四十五年(1706年),康熙途经宽河城,正值板栗成熟,食后赞曰:"天下美味也。"

"家家栽植板栗树,村村板栗树成行。"这是宽城这个"中国板栗之乡"的真实写照。目前,宽城全县板栗种植面积达80万亩,有栗树2600万株,百年以上的板栗古树达10万余株,现存最老的板栗古树树龄近700年。河北宽城传统板栗栽培系统就是以万株百年以上树龄的板栗林为特色的板栗复合栽培系统。板栗不仅满足了宽城人的口腹,还深刻影响着当地人的生活:作为当地的农业主导产业,板栗带来的经济收益,占到宽城农业收入的80%以上。

在讲求速度和效率的今天,宽城一直坚持传统的板栗种植方法。宽城林地面积有180万亩,森林覆盖率达到62%。当地农民根据地形、气候等自然条件的差异,在不同地段进行不同植被的合理搭配:修建撩壕和梯田,栽植板栗;在林下种植低矮农作物,饲养家禽;用剪下的栗树枝条种植栗蘑;利用物理和生物方法防治病、虫、草害。在机械和农药还没有发明出来的日子里,梯田—板栗—作物—家禽的复合生产体系,在宽城延续了一代又一代。这种种植模式不仅遵循了自然生态的规律,还构成了独具特色的山地林农景观,发挥着水土保持和水源涵养的重要作用。

地理位置:北纬40°17′—40°45′、东经118°10′—119°10′,地处河北省承德市宽城满族自治县。
气候特点:温带大陆性季风气候。
认定时间:2014年被农业部列为第二批中国重要农业文化遗产。

Geographical location: 40°17′-40°45′N, 118°10′-119°10′E; located in Kuancheng Manchu Autonomous County, Chengde City, Hebei Province.
Climate type: temperate continental monsoon climate.
Time of identification: In 2014, it was listed as one of the second batch of the China-NIAHS by the Ministry of Agriculture.

Typical Intercropping Mode of Forest and Crops in North China
Kuancheng Traditional Chestnut Cultivation System

Braised Pork with Chestnuts is a home-cooked dish in Jiangsu and Zhejiang provinces. The main material of this dish—chestnuts—however, becomes a recreational snack after people have a meal in North China. Whether it's tumbling in the black iron pots on the street alleys or in the exquisite snack cabinets in a mall, the sweet taste of the chestnuts that has been sizzled and baked will always give people a hint of comfort in the chilly midwinter.

Chestnut trees have been planted for thousands of years in China. As the birthplace of chestnuts, China currently holds more than 300 chestnut varieties. Nowadays, most of China's provinces such as Liaoning, Inner Mongolia, Jiangsu, Shandong, and Shaanxi can see the traces of chestnut trees, and the annual output of chestnuts ranks first in the world. Among them, the chestnuts produced in the Yanshan Mountains are in superior quality. Kuancheng County, located in the Yanshan Mountains, with high iron content, abundant sunshine and large temperature difference between day and night, is very suitable for chestnut growth. Chestnuts have been planted for more than 2000 years here. It is rumored that in the 45th years during the reign of Emperor Kangxi (1706), the chestnuts were in the ripening season when Kangxi passed by the Kuancheng County. And after tasting, he praised: Amazing delicacy!

"You cannot live in a household without chestnuts and cannot step into an alley without them, either." As the poem described, the scene is a true portrayal of Kuancheng County— "The Home of Chestnuts". By now, the area of chestnut planting in Kuancheng County is 800,000 mu. There are 26 million chestnut trees and over 100,000 chestnut trees having been in existence for more than 100 years. The oldest chestnut tree is nearly 700 years old. The traditional chestnut cultivation system in Kuancheng County belongs to a compound cultivation system featuring tens of thousands of more than 100 years old chestnuts tress. Chestnuts not only feed the Kuancheng people, but also play an important role in their life: As the leading agricultural industry in the county, the economic benefits brought by chestnut account for more than 80% of the income from Kuancheng agriculture.

Though pushed by today's high pace of life, people in Kuancheng have always adhered to traditional methods of chestnut cultivation. The area of Kuancheng forest land is 1.8 million mu, with 62% of the forest coverage rate. In accordance with the differences in natural conditions such as topography and climate, local farmers carry out irrational arrangement of different vegetation in different area: plant chestnuts in trenches and terraces; plant low-lying crops and raise poultry under the trees; cultivate chestnut mushrooms with chestnut branches cut off; use physical and biological methods to control diseases, insects and weeds. In the days when machinery and pesticides have not been invented, the compound terrace-chestnut-crop-poultry production system has continued for generations and generations in Kuancheng County. This planting mode not only follows the laws of natural ecology, but also constitutes a unique mountain forest-agricultural landscape, which plays an important role in soil and water conservation.

宽城板栗栽培可追溯至东汉时期，至今已近 2000 年。时至今日，全县板栗种植面积达 80 万亩。图为生长在著名的喜峰口长城脚下的宽城板栗，它们共同见证了伟大的中华文明。（郝振平 摄）

The chestnut cultivation in Kuancheng County can be traced back to the East Han Dynasty, which has been more than 2,000 years since then. Today, the county's chestnut planting area is 800,000 mu. Chestnuts growing at the foot of the famous Great Wall at Xifengkou, have witnessed the great Chinese civilization. Photographed by Hao Zhenping

宽城县地形以山地为主，农民根据不同地段地形、土壤、气候等自然条件的差异，进行不同植被的合理搭配，构成了独具特色的山地森林景观。其中，山坡的中下部土层较厚、肥沃，土壤水分条件好，适宜栽植板栗。（上图 胡九龙 摄；下图 郝振平 摄）

Kuancheng County is mainly in mountainous. Farmers carry out reasonable arrangement of different vegetation according to the differences of natural conditions such as topography, soil and climate in different sections, which constitute a unique mountain forest-agricultural landscape. The middle and lower parts of the hillside, with good soil moisture conditions, are thick and fertile and suitable for planting chestnut trees. The picture above photographed by Hu Jiulong, and below by Hao Zhenping

宽城传统板栗栽培系统是一种可持续的生态农业生产模式。传统板栗园利用物理和生物方法防治病、虫、草害，形成梯田—板栗—作物—家禽复合生产体系。图为栗树与大豆、谷子、栗蘑以及林下养殖相符合的生态农业景观。（闵庆文 摄）

The traditional chestnut cultivation system in Kuancheng County is a sustainable and environment friendly agricultural production model. Physical and biological methods are used for controlling diseases, insects and weeds, and then a compound terrace-chest-crop-poultry production system is formed. The pictures show the eco-agricultural landscape in which chestnut trees are planted with soybeans, millet, chestnut mushrooms and poultries. Photographed by Min Qingwen

宽城地处燕山深处，土壤富含铁、光照充足、雨热同期、昼夜温差大，适宜板栗生长。每年的 5—6 月份，板栗开花，不久以后一颗颗碧绿的刺球状硕果便挂满枝头，青翠欲滴。9—10 月份是板栗成熟和采摘、采购的季节，彼时全城栗香。（胡九龙、郝振平 摄）

Kuancheng County, surrounded by the Yanshan Mountains, with rich iron in soil, sufficient sunshine, simultaneous rain and heat, and large temperature difference between day and night, thus is suitable for chestnut growth. Every year from May to June, chestnut blossoms, and soon after, branches will be covered by aquamarine green thorn-like fruits. September to October is the seasons of ripening, picking and purchasing of chestnut, and the city will be full of chestnut fragrance. Photographed by Hu Jiulong and Hao Zhenping

宽城板栗果面褐红油光，果肉黄白适宜，果实饱满，粒大均匀，食用甘甜适口，有"东方珍珠"的美誉。板栗栽培自古以来就是宽城农业的主导产业。板栗被誉为"铁杆庄稼""木本粮食"，是当地居民主要的食物来源之一。图为板栗丰收，以及板栗的初加工和深加工。（郝振平 摄）

The brownish and oily chestnut fruit surface, the yellow and white flesh, the full fruit, the even figure and the sweet and palatable tasty, the chestnuts in Kuancheng County deserve the reputation of "Oriental Pearl". Chestnut cultivation has been the leading industry of Kuancheng agriculture since ancient times. Known as "iron set crops" and "woody food", chestnut is one of the main food sources for local residents. The pictures show a scene of harvest, the initial processing and deep processing of the chestnuts. Photographed by Hao Zhenping

当地人将板栗看作吉祥的象征，在拜师、庆寿、婚嫁等重要时刻，都以栗子相赠，以示祝福。有关板栗的历史传说、民俗礼仪、文学作品不胜枚举，展现出丰富多彩的板栗文化。图为2016年9月，当地举办首届板栗采摘节。（郝振平 摄）

The locals regard chestnut as a symbol of good fortune. They give their loved chestnuts at important moments such as apprenticeship, birthday, and marriage day to show their blessings. The history, folk etiquette, and literary works of the chestnut are numerous, which show a rich and colorful chestnut culture. The picture shows The First Chestnut Picking Festival held in September, 2016. Photographed by Hao Zhenping

北方梯田与石头文化的集成
河北涉县旱作梯田系统

早在秦汉时期，中国就出现了梯田。生活在南方地区的农民，为了解决粮食问题，在丘陵上，沿着等高线建起了一道道堤坝。梯田的出现，为生活在南方的人们提供了涵养水源的方法，人与水的"战争"完美解决。然而，在年降水量仅540毫米的河北涉县，修梯田更像是人与石头的"战争"。

地处太行山东麓石灰岩深山区的涉县境内均为山地。这里石压着石，山叠着山，可供种植的土地中，土层最厚处为0.5米，最薄处仅为0.2米。"山高坡陡、石厚土薄"是这里的真实写照。生活在这里的人们，用修建梯田的方式，与大自然做抗争。据考证，从元代起，就有人在涉县修建梯田。到了清代，社会安定、人口增长速度加快，人们对粮食的需求增加，梯田开始初具规模。这里梯田规模最大的是井店镇王金庄，梯田面积达1.2万亩，石堰长度近万华里，高低落差近500米。

从山脚到山顶，农民们顺坡而上，根据不同的山势进行梯田的开垦和作物的种植：山顶栽松柏；山腰梯田中种植谷子和玉米；田埂种植花椒、黑枣、柿子；路边石缝间嫁接枣树；在耕地旁栽植泡桐。河北涉县旱作梯田系统就是以北方山地种植谷物、花椒为特色的旱作梯田复合生态系统，该系统在解决了基本粮食需求的基础上，增加了作物的多样性。

恶劣的自然环境激发了人们进行不断创造的智慧。梯田中随处可见集雨水窖，供农民休息避雨的石屋也散落其中。他们还发明了一种"悬空拱券镶嵌"式的石堰修复技术：在当地就地取材，把石头一片叠一片，层层叠加，不需要水泥、石灰等黏合剂。采用这种技术修的堤坝不仅能够防止因山高坡陡、洪水成灾造成的梯田坍塌，还能防止水土流失，蓄水保墒。毛驴的使用，则更加凸显了人的智慧——它们同时承担着生产工具、运输工具和有机物转化的重要工作。石头、作物、梯田、毛驴，在人的作用下巧妙结合，相得益彰，融合为五位一体的可持续发展的旱作农业生态系统。

地理位置：北纬36°17′—36°55′、东经113°26′—114°，地处河北省邯郸市涉县。
气候特点：温带大陆性季风气候。
认定时间：2014年被农业部列为第二批中国重要农业文化遗产。

Geographical location: 36°17′-36°55′N, 113°26′-114°E; located in She County, Handan City, Hebei Province.
Climate type: temperate continental monsoon climate.
Time of identification: In 2014, it was listed as one of the second batch of the China-NIAHS by the Ministry of Agriculture.

Integration of Terraced Fields and Stone Culture in North China
Shexian Dryland Farming Terraces System

As early as the Qin and Han Dynasties, there were terraces in China. In order to solve hunger, the peasants living in the southern region built dams along the contour line on the hills. The emergence of terraces provided a way for people living in the south to conserve water, and the "war" between people and water was perfectly settled. In She County, However, where the annual precipitation is only 540 mm, constructing terraces is more like a "war" between people and stones.

She County is seated at the deep limestone mountainous areas, east of Taihang Mountains. Stones and mountains are everywhere, and among the land that can be planted, the thickest part of the soil is 0.5 meters, and the thinnest is only 0.2 meters. "High mountains, steep hills while thick stones and thin soil" is a true portrayal of this place. People living here use the way of building terraces to fight against natural environment. It is recorded that people began to build terraces in She County from the Yuan Dynasty. In the Qing Dynasty, social stability accelerated population growth, people's demand for food increased, and then terraces began to take shape. The largest terraced area here is in Wangjinzhuang Village, Jingdian Township. The terraced area is 12,000 mu, and the total length of the stone is nearly 5000 kilometers. The height difference is nearly 500 meters.

From the foot to the top of the mountain, the farmers take advantages of the slope and assart terraces and plant crops based on different terrain: Plant pines and cypresses on the top of the mountain; plant millet and corn in the terraced fields; plant Chinese pepper, black jujubes and persimmons on the edge of the mountainside; graft jujube trees on the roadside and between narrow stone crevices; plant paulownia next to the cultivated land. Dryland Farming Terraces System in She County is a compound ecosystem featuring cereals and Chinese pepper, which increases crop diversity on the basis of addressing basic food needs.

The hard-natural environment inspires the ever-creating wisdom. Rainfall cisterns can be seen everywhere in the terraces, and stone houses for farmers to rest and shelter from the rain are also scattered. A "hanging vault inlaid" technique is also used for building and repairing stone dams: stones from the surroundings are stacked one on another, no cement, lime and other adhesives are needed. This technically repaired dam can not only prevent the collapse of terraces caused by steep slopes and floods, but also prevent soil erosion, and then water can be preserved. The use of donkeys highlights the wisdom of human beings. They play the roles as farm tools, transportation "vehicles" and can produce organic manure. Stones, crops, terraces, and donkeys are combined under the influence of human beings to coordinate with each other and become a five-in-one sustainable dryland farming ecosystem.

河北涉县旱作梯田系统位于河北省西南部的晋、冀、豫三省交界处，地处太行山东麓。涉县境内均为山地，全县旱作梯田总面积达 21 万亩。图为冬日里的涉县旱作梯田。（赵仁义 摄）

The Dryland Farming Terraces System is located in the southwestern part of Hebei Province, at the junction of three provinces of Shanxi, Hebei and Henan, and is in the eastern foothills of Taihang Mountains. The total area of dry farming terraces in the mountainous county is 210,000 mu. The picture shows the dryland farming terraces in She County in the winter. Photographed by Zhao Renyi

与许多南方梯田不同，涉县旱作梯田以石头修葺田埂，展现了人工与自然的巧妙结合。在山巅登高望远，用石头垒起的梯田，犹如一条条巨龙蜿蜒起伏于座座山谷之间，并随着季节的变化呈现出各种姿态。（上图 赵仁义 摄；下图 黄亚军 摄）

Unlike many southern terraces, the dryland farming terraces in She County are built with stone to consolidate the terrace ridges, which shows the ingenious combination of man and nature. See from the mountain top, the terraces that are built with stones are like a dragon, undulating in the valleys, and showing various postures as the seasons change. The picture above photographed by Zhao Renyi, and below by Huang Yajun

梯田里农林作物丰富多样，谷子、玉米、花椒、柿子、黑枣等漫山遍野，各类瓜果点缀在万亩梯田里，呈现春华秋实的壮丽景象，迸发出人与自然的和谐之美。（黄亚军 摄）

There are rich and varied agricultural and forestry crops in the terraces. Millet, corn, Chinese pepper, persimmon, black jujubes and the like are all over the mountains. A variety of fruits are dotted in the vast terraces, showing a magnificent harvest scene and a brilliant harmonious beauty between man and nature. Photographed by Huang Yajun

春夏时节，当地农人都会在田间耕种、除草。除石砌的田埂外，农民往往搭建石屋，用于避雨、午餐和休息等。（黄亚军 摄）

In the spring and summer, local farmers will cultivate and weed in the fields. In addition to the terrace ridges, farmers often build stone houses to have a lunch or rest or shelter from rain. Photographed by Huang Yajun.

和南方梯田一样，涉县旱作梯田石砌田埂也根据山势陡峭情况而有高低之分，缓坡的石砌田埂较矮，陡坡的田埂则修得很高，有的达到几米的高度。为此，当地人发明了一种"悬空拱券镶嵌"式的石堰修复技术，对石堰进行加高、加固，以最大限度地保土保墒。即便如此，也难免会有石堰垮塌的情况出现。按照近年来涉县人民政府出台的《涉县旱作梯田修复建设及保护发展实施方案》，当地农民每年都会对石堰做相应的检查和修复。（黄亚军 摄）

Like the terraces in South China, the dryland farming stone terraces in She County also vary in height based on the steepness of the mountains. The stone terraces on gentle slopes are relatively short, while the one on an abrupt slope can reach a height of several meters. Then to heighten and reinforce the terrace ridges, the "hanging vault inlaid" technique is created, which maximizes soil and water conservation. Even so, it is inevitable that there will be stone collapses. According to *the Implementation Plan of Construction and Protection Development for the Dryland-farming Terraces in She County* issued by the people's government of She County in recent years, the locals will carry out recondition for the stone ridges every year. Photographed by Huang Yajun

目前涉县正对旱作梯田进行合理科学修复，并组织开展农田道路、主干排水渠、抗旱水窖、蓄水塘坝等基础设施建设，逐渐形成"小雨润物、中雨蓄墒、大雨入塘、暴雨进川、水不出山"的良性生态体系。图为当地农民带着毛驴和锄头到田间耕作，以及打造山间集雨水窖的情景。（黄亚军 摄）

At present, She County is carrying out reasonable and scientific restoration of the dryland farming terraces, and building infrastructure such as farmland roads, main drainage channels, drought-resistant cisterns, and storage ponds and the like. And gradually, a benign ecological system that the drizzles can moisten the plants, moderate rains can be stayed the water in the soil, heavy rains can be stored in the ponds, intense falls can be drained away to the rivers, and requisite water can be kept in the area is formed. The pictures show local farmers with donkeys and hoes to the field to do farm work and built a cistern. Photographed by Huang Yajun

除了毛驴、石屋，涉县旱作梯田系统不得不提的还有花椒树。在梯田堰边种植花椒树，既可增加经济收入，又可作为生物埂保护梯田，这是王金庄梯田作物与花椒间作套种种植模式，而花椒的采摘时间一般在立秋前后。（贺献林 摄）

In addition to the donkeys and stone houses, the Chinese pepper trees in She County's dryland farming terrace system are also worth to being mentioned here. Planting Chinese pepper trees on the terraces side can help increase income and protect the terraces as a bio-bank. This is the intercropping crop-Chinese pepper terrace mode in Wangjinzhuang Village. The picking time of pepper is generally before and after the Beginning of Autumn. Photographed by He Xianlin

世界旱作农业源头
内蒙古敖汉旱作农业系统

一路向北，越过了秦岭—淮河线，耕地大多从水田转向了旱地。人们依靠天然降水进行农业生产。这便被称为"旱作农业"。同时，人们种植的作物也由对水资源要求较高的水稻，向更耐干旱的小麦、粟黍转换。

粟黍，被认为是北方旱作农业的代表作物。有资料表明，中国北方进行旱作农业比中欧地区早了 2000 多年，而这一判断有着有力的证明——在内蒙古敖汉旗的兴隆沟遗址，曾经出土过一批炭化谷物，被认为距今已有 7700 年到 8000 年的历史，比中欧地区发现的谷物早 2700 多年。这是中国北方旱作农业谷物的重要实证，也正是因为这批谷物，兴隆沟遗址被认为是"旱作农业发源地"。在兴隆洼文化遗址（兴隆沟遗址之一）出土的文物中，还有大量的石器、骨器等，其中石杵、石斧、石铲、石刀等，则大多为原始农耕生产用具。

敖汉旗，位于今内蒙古赤峰市，地处农耕区和游牧区交错的地带，是农业文明与草原文明的交汇处。在这片科尔沁沙地南缘的土地上，有着超过 3000 处古文化遗存，曾出土过大量的石器、骨器等原始农耕生产用具，证实早在 1 万多年前，就有人类在这里生息繁衍。悠久的种植历史，让人们对敖汉旗农作物品种的丰富多彩感到毫不意外——光是粟就分了四种颜色。这里远离现代农业技术，地理环境和自然风貌与原始时代相比没有大的改变，原始而又朴素的农业种植形态被保留了下来，谷子、糜黍、荞麦、高粱、杂豆等杂粮成为当地的优势产业。这个年均降水量不足 500 毫米的地方，竟是全国的粮食生产先进县。敖汉旱作农业系统就是以旱地种植谷子（小米）为主，兼顾其他杂粮种植为特色的旱作农业生态系统。

在数千年的发展中，生活在这里的人们和他们种植的作物一起协同进化，适应着当地的自然环境，并形成了一系列独有的民间文化。正月初八祭星是敖汉人所独有的祭祀风尚，此习俗至今在四家子镇牛汐河屯仍在保留延续。位于敖汉旗境内的国家级重点文物保护单位城子山遗址，被专家称为"中国北方最大的祭祀中心"，还有诸多不同时期的出土文物，均与祭祀有关。

地理位置：北纬 41°42′—43°02′、东经 119°30′—120°53′，地处内蒙古自治区赤峰市敖汉旗。
气候特点：温带大陆性季风气候。
认定时间：2012 年被联合国粮农组织列为全球重要农业文化遗产，2013 年被农业部列为首批中国重要农业文化遗产。

Geographical location: 41°42′-43°02′N, 119°30′-120°53′E; located in Aohan Banner, Chifeng City, Inner Mongolia Autonomous Region.
Climate type: temperate continental monsoon climate.
Time of identification: In 2012, it was listed as one of the GIAHS by the FAO. In 2013, it was listed as one of the first batch of the China-NIAHS by the Ministry of Agriculture.

The Birthplace of the World's Dryland Farming

Aohan Dryland Farming System

All the way to the north, crossing the Qinling-Huaihe line, the water resources begin to become scarce, and the cultivated land turns from paddy fields to dry land. People rely on natural precipitation for agricultural production. At the same time, the crops that people tend to plant on large scales are switching from rice with high water requirements to drought-tolerant wheat and millet.

Millets is considered to be the representative crops of dryland farming in the north. According to some documentations, people in North China began to have dryland farming more than 2,000 years earlier than people in Central Europe did. This judgment has a strong proof: a number of charred grains have been unearthed at the Xinglonggou site in Aohan Banner, Inner Mongolia. It is believed to have a history of 7,700 to 8,000 years, 2,700 years earlier than the grains found in Central Europe. This is the important demonstration of dryland farming cereals in northern China. It is because of the grains that the Xinglonggou site is considered to be the "Birthplace of Dryland Farming". In the cultural relics unearthed from the Xinglongwa Site (one of Xinglonggou site), there are also a large number of stone tools, bone tools and the like. Most of them are primitive farming tools, such as stone hoes, stone axes, stone shovel, stone knives, etc.

Aohan Banner, Chifeng, Inner Mongolia, is located at the intersection of agricultural and nomadic areas, and is also the intersection of agricultural civilization and grassland civilization. On the land at southern foot of the Horqin Sandy Land, there are more than 3,000 ancient cultural relics. A large number of primitive agricultural production tools such as stone tools and bone tools have been unearthed. It is confirmed that humans lived here more than 10,000 years ago.

People won't feel surprised about the abundant varieties of crops in Aohan Banner since they have enjoyed a long history: just the millet alone has four kinds of color. Modern agricultural technology is not available here, and the geographical environment and natural features have not changed much compared with the original. Hence, the original and simple agricultural planting pattern has been preserved. The foxtail millet, the broomcorn millet, the buckwheat, the sorghum, the miscellaneous beans and other miscellaneous grains have become local advantageous industries. The place, where the average annual precipitation is less than 500 mm, is unexpectedly an advanced county for grain production in China. The Aohan Dryland Farming System is actually an agro-ecosystem characterized by planting the foxtail millet (also the millet) in dryland and taking other miscellaneous grains into account.

With the development of thousands of years, people living here and their crops have evolved together to adapt to the local natural environment and form a unique folk culture. A sacrificial ceremony called Jixing (a day all gods of stars descend to the world and people light candles to pray for peace and good fortune) on the fifteenth day of the first month of lunar year is a unique custom in Aohan, which has been remained and inherited in Niuxihetun Village, Sijiazi Township. The Chengzishan site, a national key cultural relic protection unit located in Aohan Banner, is identified by experts as the largest ritual center in northern China. Besides, there are many unearthed cultural relics from different periods, which are all related to sacrificial ceremony.

敖汉旗位于燕山山脉东段北麓，科尔沁沙地南缘，这里山川秀美，沃野无边，是世界旱作农业的发源地。独特的地理环境和气候条件为这里旱作农业的发展提供了基础，坡坡岭岭，沟沟坎坎，到处都是以粟、黍和高粱等为代表的杂粮作物。（于海永 摄）

Aohan Banner, located at the eastern section of northern Yanshan Mountains, and the south of Horqin Sandy Land, with its boundless fertile fields, is the birthplace of the world's dryland farming agriculture. The unique geographical environment and climatic conditions provide the basis for the development of dryland farming here. The panicum, the broomcorn millet and sorghums are planted on every slopes and hills. Photographed by Yu Haiyong

敖汉旗历史文化悠久，尤其是以当地地名命名的小河西、兴隆洼、赵宝沟、小河沿四种史前文化，以其丰富的遗存，填补了中国北方考古编年的空白。有专家推断，这里是世界旱作农耕文明的源头。图为敖汉旗四家子镇扣合林村沐浴在晨光秋色之中。（于海永 摄）

Aohan Banner is famous for its long history and culture, especially the four prehistoric cultures: Xiaohexi, Xinglongwa, Zhaobaogou and Xiaoheyan named after the local names. The rich remains fill the blank of archaeological chronicles in northern China. Some experts have concluded that Aohan Banner is the birthplace of the world's dryland farming civilization. Kouhelin Village in Sijiazi Township is bathing in the morning light of the autumn in the picture. Photographed by Yu Haiyong

敖汉旗的粟和黍是原始的栽培物种，在当地的区域分布和种植季节上具有互补性和不可替代性，并且生育期短，适应性强，耐旱耐贫瘠，是干旱、半干旱地区发展旱作节水农业的重要作物。图为成熟的大红黍。（于海永 摄）

The panicum and the broomcorn millets are native crop varieties in Aohan Banner, which is complementary and irreplaceable for local regional distribution and planting seasons. They have short growth period, strong adaptability, good drought and leanness tolerance. Hence, they are prior choices for developing dryland and water-saving agriculture in arid and semi-arid areas. The picture shows the ripen red broomcorn millet. Photographed by Yu Haiyong

作为典型的旱作农业区，敖汉旗杂粮生产是其优势产业，其中谷子是第一大杂粮作物。该地杂粮绝大部分种植在山地或沙地上，自然条件较好，极少使用化肥农药，保证了杂粮生产的天然特性。图为成片的谷子地和其间的辽代古塔沐浴在秋风中。（于海永 摄）

As a typical dryland farming area, the production of miscellaneous grains in Aohan Banner is its highly competitive industry, and the foxtail millet is in a dominant position. Most of the local miscellaneous grains are planted in mountainous or sandy land. With good natural conditions and few uses of chemical fertilizers and pesticides, the natural characteristics of the grains can be ensured. The picture shows the vast millet field and the ancient tower is standing in the autumn wind. Photographed by Yu Haiyong

敖汉旗的农作物品种丰富多样，最有名的粟有黑、白、黄、绿四种颜色。黍的品种也很多，有大粒黄、大支黄、大白黍、小白黍、疙瘩黍、高粱黍和庄河黍等。上图为当地农民在扬场，借助风力使黍壳与黍粒分离；下图为农民在收获谷子。（邹宝良 摄）

Aohan Banner has a wide variety of crops, in which the most famous millet panicum has four colors: black, white, yellow and green. There are also many varieties of the broomcorn millet, such as yellow broomcorn millet with big gains, white broomcorn millet with big grains, white broomcorn millet with small grains, lump broomcorn millet, sorghum millet and Zhuanghe broomcorn millet. The pictures show local farmers are separating the shells from the grains with the help of the wind or locals are harvesting. Photographed by Zou Baoliang

除了粟和黍，敖汉旱作农业系统中还有其他很多粮食作物、经济作物、蔬菜、瓜果和畜禽等。以上图片为当地农民在土豆、蘑菇等蔬菜田间劳作、收获。（邹宝良 摄）

In addition to the panicum and the broomcorn millet, there are many other food crops, cash crops, vegetables, fruits and livestock in the Aohan Dryland Farming System. The pictures show that the locals are busy with harvesting. Photographed by Zou Baoliang

华北地区 · North China / 内蒙古自治区 · Inner Mongolia Autonomous Region

荞麦也是敖汉旗的重要农作物。这是一种一年生草本植物，花白色或粉红色。荞麦喜凉爽湿润，具有丰富的营养价值。图为开满花的荞麦田。（邹宝良 摄）

Buckwheat is also an important crop in Aohan Banner. This chimonophilous crop, planted in humid environment, is an annual herb with white or pink flowers. It has high edible value. Buckwheat flowers bloom around the field. Photographed by Zou Baoliang

流传在敖汉旗境内的庙会、祭星、祈雨、撒灯等民俗以及民间的扭秧歌、踩高跷、唱大戏等，大都是为了祈求一年风调雨顺、五谷丰登和庆祝丰收。图为当地农民元宵节撒龙灯转庙请圣火和踩高跷的场景。（于海永 摄）

The temple fairs, Jixing (which has been mentioned above), the lantern festival, the lights, as well as yangko dance, stilts, and operas that have been survived in the Aohan Banner are mostly for praying and celebrating for a good weather and yield. Local farmers are delivering the holy flame temple by temple, lighting dragon lanterns in the Lantern Festival and walking on stilts in the Lantern Festival. Photographed by Yu Haiyong

敖汉旗中南部地区，自清乾嘉时期开始，在元宵节期间还会举办"黄河灯会"，俗称"跑黄河"或"转九曲"。其形式类似走迷宫，参与者要从头跑到尾。图为踩高跷的民众在"跑黄河"。（于海永 摄）

In the central and southern areas of Aohan Banner, "Yellow River Lantern Festival", commonly known as "Running along the Yellow River" or "Zhuan Jiuqu" has been hold during the Lantern Festival since the Qing Dynasty reigned by Emperor Qianlong. The participants have to run from the beginning to the end of a "labyrinth". The picture shows the people on the stilts are enjoying the game. Photographed by Yu Haiyong

蒙古族游牧文化的缩影
内蒙古阿鲁科尔沁草原游牧系统

位于北纬44°线的南北两侧、大兴安岭西麓的中国大陆内部、蒙古高原的边缘，这片海洋季风在中国国土范围内最后到达的土地，年均降水量不足400毫米，木本植物难以生长。土壤层薄，气候高寒、干旱，传统的农业耕种在这里不被支持。但是，这里却孕育了连片的草原。它们被用蒙古族语言赋予了有象征意义的名字：呼伦贝尔、锡林郭勒、科尔沁……

生活在草原上的人们，以放牧为生，被称作"牧民"。他们逐水草而居，来获取生活资料，形成了牧民—牲畜—草原（河流）的三角依存关系。这种生活方式被称作"游牧"，距今已有3000年的历史。

牧民们深谙取物有时的道理。他们熟知当地河流、草场的季节变化，根据当年雨水和草场长势，来决定一年的游牧线路以及春、夏、秋、冬四季牧场的放牧时间。这种生活方式使人和牲畜不断迁徙、流动，保证了牧群不间断获得充足饲草的同时，又能够避免对一片草地的过度使用。这也是保持草场能够持续被利用的最佳方式。这不断孕育和发展着的蒙古族人民的生产方式、生活习俗、文化特质和宗教信仰，时刻体现着深藏在蒙古族人民血脉之中的崇尚天意、敬畏自然、天人合一的生活理念。

科尔沁草原和锡林郭勒草原的交接地区，有一片历史悠久的天然牧场——阿鲁科尔沁草原。阿鲁科尔沁草原位于今赤峰市。这里是理想的游牧场地：河流密布、草原广袤——核心区超过4000平方千米，使游牧活动全年都能有充足的水草资源。在此之外，大兴安岭既为它阻挡了来自西伯利亚的寒流，还给生活在这里的牧民提供了制造工具的木材；这里还与农耕区接壤，牧民可以方便地获得牧区无法生产制造的用品。

在讲求"智能"的当下，这里仍保留着以牛、羊、马为主要牲畜品种的传统草原游牧系统。牧民—牲畜—草原（河流）的三角依存关系在阿鲁科尔沁草原被延续至今。蒙古族传统的生产生活方式、文化特质和宗教信仰在这片草原上被完整地保留了下来，并在牧民们中间代代传承。

地理位置：北纬43°21′—45°24′、东经119°02′—121°01′，地处内蒙古自治区赤峰市阿鲁科尔沁旗。

气候特点：温带大陆性季风气候。

认定时间：2014年被农业部列为第二批中国重要农业文化遗产。

Geographical location: 43°21′-45°24′N, 119°02′-121°01′E; located in Aru Horqin Banner, Chifeng City, Inner Mongolia Autonomous Region.

Climate type: temperate continental monsoon climate.

Time of identification: In 2014, it was listed as one of the second batch of the China-NIAHS by the Ministry of Agriculture.

The Epitome of Mongolian Nomadic Culture
Aru Horqin Grassland Nomadic System

The inland, located on the north and south sides of the 44° north latitude and the northern foot of the Greater Khingan Range Mountain, is influenced by the marine monsoon, resulting in the average annual precipitation less than 400 mm and a hard environment difficult for woody plants to grow. With thin soil layer, cold and dry weather, traditional agricultural farming cannot be carried out here. However, it is the area that breeds the boundless grasslands which are named in Mongolian language as Hulun Buir Grassland, Xilingol Grassland, Horqin Grassland...

People there make a living by grazing and are called herdsmen. They migrate to where water and grass are available and form a triangular dependence relationship among herdsmen, livestock and grasslands (rivers). The nomadic lifestyle has a history of 3,000 years.

The herdsmen are quite familiar with the principle: different season, different plan. They determine the nomadic route and the grazing time of the year based on their knowledge of the seasonal changes of local rivers and pastures. This lifestyle allows people and livestock to migrate continually, which ensures that the herds can receive sufficient forages and avoids excessive use of one grassland at the same time. This is also the best way to keep sustainable pastures. The unique production methods, living customs, cultural traits and religious beliefs, which are constantly nurtured and developed, always embody the people's admiration for nature and nature unity which are deeply hidden in the blood of the Mongolian people.

In the junction of Xilingol Grassland and Horqin Grassland, there is a time-honored natural pasture in Chifeng—the Aru Horqin Grassland. The core area, with a dense network of rivers and vast grasslands, is over 4,000 square kilometers, allowing nomadic activities to have sufficient resources throughout the year. Beyond that, the Greater Khingan Mountain not only blocks the cold wave from Siberia, but also provides the herdsmen living here with the woods to make tools. Adjacent to the farming area, herdsmen can easily obtain articles of daily use that cannot be produced in the pasturing area.

In the age of artificial intelligence, the grassland there still retains a traditional nomadic grassland system with cattle, sheep and horses as the main livestock species and the triangular dependence relationship among herdsmen, livestock and grasslands has continued in the Aru Horqin Grassland to this day. The unique production method, lifestyle, cultural traits and religious beliefs of the Mongolian people have been preserved on this grassland and passed down from generation to generation.

内蒙古阿鲁科尔沁草原游牧系统，是以牧民—牲畜—草原—河流相互依存、和谐共生为特色的复合游牧生态系统，核心区位于今内蒙古赤峰市阿鲁科尔沁旗巴彦温都尔苏木，面积4141平方千米，自古以来就是游牧民族狩猎的场所和游牧活动的栖息地。（邹宝良 摄）

The nomadic grassland system in Aru Horqin Grassland, Inner Mongolia is a compound nomadic ecosystem characterized by the interdependence and harmonious symbiosis among the herdsmen, livestock, grasslands and rivers. The core area in Bayan Wendur Sumu, Aru Horqin Banner, Chifeng, with an area of 4,141 square kilometers, has been a habitat for hunting and nomadic activities since ancient times. Photographed by Zou Baoliang

阿鲁科尔沁草原游牧系统长期演化的历史过程和现实存在，向人们阐释了一个取物有时的道理。在农耕化浪潮和现代农牧业技术出现之前，对于生活在科尔沁草原上的历代游牧民来说，"逐水草而居"是唯一可行的生产生活方式。图为阿鲁科尔沁旗巴彦温都尔苏木的牧民在转场途中（上图）和牧民在为长了寄生虫的羊喷洒药物（下图）。（胡国志 摄）

The historical process and reality of the long-term evolution of the nomadic system of Aru Horqin Grassland has explained to people that we must take from the nature according to the time. Before the emergence of mechanization of farming and the modern technology of farming and animal husbandry, migrating to where water and grass are available is the only viable production lifestyle for the nomads living on the Horqin Grassland. In the pictures, nomads are on a migration and spraying insecticide for sheep that have grown parasites. Photographed by Hu Guozhi

蒙古族的祭祀与游牧生活息息相关，这其中最隆重的是祭敖包。人们通过祭敖包祈求天地神保佑人间风调雨顺、牛羊兴旺、国泰民安。图为阿鲁科尔沁旗在举行敖包祭祀仪式，人们奉上哈达、祭品，焚香点火、诵经、跪拜后，围绕敖包顺时针方向转三圈，祈求降福，保佑人畜两旺。（永平 摄）

The Mongolian rituals are closely related to the nomadic life, in which the most important is the ritual for Obo. By worshiping the Obo, people pray for that wind and rain come on time, the cattle and the sheep are all thrive, the nationality is prosperous and the people are at peace. The pictures show that people in Aru Horqin Banner are performing the ritual. They have the honour to send the Khata and sacrificial offerings. After burning incense, chanting sutras and kowtowing, they turn around the Obo clockwise three times, paying for good fortunes. Photographed by Yong Ping

每年七八月牲畜肥壮的季节，正是蒙古族举行那达慕大会的日子。"那达慕"，蒙语的意思是娱乐或游戏，是人们为了庆祝丰收而举行的文体娱乐大会。那达慕大会上有惊险刺激的赛马、摔跤，令人赞赏的射箭，争强斗胜的棋艺，引人入胜的歌舞等活动。（永平 摄）

Each year in July and August when the livestock are stout and strong, the Nadam fair of the Mongolian people will be held. Nadam, which means entertainment or games, is a cultural and athletic entertainment conference held to celebrate the harvest. At the Nadam fair, there are thrilling horse racing, wrestling, admirable archery, chess games, and fascinating songs and dances. Photographed by Yong Ping

农历腊月二十三，是我国北方地区一年一度的小年，而对于蒙古族同胞来说，这一天则是非常重要的祭火节，它与游牧民族的生产生活有着千丝万缕的联系。祭火的仪式非常隆重，人们往往在一两天前就开始打扫庭院、房屋，准备祭品，搭建火撑子。祭火开始，全体参加者需要跪拜，祈求人丁兴旺、五畜昌盛！（永平 摄）

It is an annual Little New Year on the 23rd day of the lunar month in the northern part of China. For the Mongolian compatriots, however, an important fire ritual inextricably linked to the nomadic production and life is held today. People often begin to clean the courtyards and houses, and prepare sacrificial offerings and a huge special furnace several days before the grand ritual. When the fire ritual begins, all participants need to bow down and then pray for their loved ones and prosperity of the livestock. Photographed by Yong Ping

由于矿产资源开发、草场过载和天然草场大量被占用，阿鲁科尔沁草原面临着生态系统恶化、生物多样性减少的威胁。同时，现代生产技术的应用和生活方式的改变，也给当地牧民传统的生产生活方式带来了巨大冲击。（邹宝良 摄）

Due to the excessive exploitation of mineral resources, overuse of grassland and overload the natural grasslands, Aru Horqin Grassland is facing the threat of ecosystem degradation and biodiversity reduction. At the same time, the application of modern production technology and the change of lifestyle have also brought great impact to the traditional production and lifestyle of local herdsmen. Photographed by Zou Baoliang

目前，阿鲁科尔沁旗已经制定了草原游牧系统的保护和发展规划，严格保护游牧系统栖息地和珍贵的草原文化遗产，使得内蒙古阿鲁科尔沁草原游牧系统不断散发出独特的魅力。（胡国志 摄）

Fortunately, the Aru Horqin Banner has drawn up protection and development plan for the nomadic grassland system. With strong protection for the habitats of the nomadic system and the precious grassland cultural heritage, the nomadic system of the Aru Horqin Grassland in Inner Mongolia can continue to exude her unique charm. Photographed by Hu Guozhi

东北地区·Northeast China

辽宁省

南果梨母株所在地——辽宁鞍山南果梨栽培系统

传统林参共作模式——辽宁宽甸柱参传统栽培体系

南果梨母株所在地
辽宁鞍山南果梨栽培系统

作为东汉末年著名文学家，位列"建安七子"的孔融，在今天，给人们留下深刻印象的不是什么名篇佳作，而是一个三岁小儿都可以讲出的小故事：孔融让梨。

在距今近 2000 年的东汉时期，梨已经作为一种寻常的水果，登上了百姓的餐桌。中国是梨属植物的发源地之一。据相关记载，梨树在中国的栽培历史，已经超过 4000 年。在长期的发展过程中，人们不断探索改善梨的品种的方法。成书于北魏末年的《齐民要术》中，就有关于梨的品种的记录，并详细描述了梨的种植技术、嫁接技术及贮藏方法等，可见当时梨树的栽培在中国的发展已相当成熟；成书于宋代的《洛阳花木记》中，记载了 27 个梨树的品种。

发展到今天，梨遍布我国河南、河北、山东、辽宁等多地，成为栽培面积和产量仅次于苹果的第二大常见水果。辽宁鞍山，这个凭借钢铁工业闻名的城市，有一种独特的梨——南果梨。

南果梨，因为是辽宁鞍山地区所独有的品种，也被叫作"鞍果"，位列中国"四大名梨"之一。鞍山南果梨栽培系统就是以种植南国梨为特色的传统栽培系统。

鞍山地处松辽平原的边缘，西北是广阔的平原，而东南的山区面积占到了全市面积的一半。温带季风性气候让这里四季分明、雨热同期、光照丰富，境内有大小河流超过 40 条，有着有利的灌溉条件。独特的生长环境，让南果梨皮薄肉厚、果肉多汁、香气浓郁。更为难得的是，南果梨的母株留存至今。

在鞍山市千山区大孤山镇对桩石村的山坡上，有一株低矮而枝蔓层叠的梨树十分惹人注目：它树下一侧已经腐烂，另一侧则枝蔓旺盛，焕发着生机，树枝上还系挂着红布条——这便是南果梨树的母株，是仅存的一株自然杂交实生苗南果梨树，自发现至今已有 150 多年的历史。

南果梨母株树上系满的红布条，代表着果农们美好的期待：希望新的一年风调雨顺、丰产丰收。而南果梨也没有让果农们失望。

地理位置：北纬 40°55′—41°12′、东经 122°49′—123°14′，地处辽宁省鞍山市千山区。
气候特点：温带季风气候。
认定时间：2013 年被农业部列为首批中国重要农业文化遗产。

Geographical location: 40°55′-41°12′N, 122°49′-123°14′E; located in Qianshan District, Anshan City, Liaoning Province.
Climate type: monsoon climate of medium latitudes.
Time of identification: In 2013, it was listed as one of the first batch of the China-NIAHS by the Ministry of Agriculture.

The Location of the Mother Plant of Nanguo Pear
Anshan Nanguo Pear Cultivation System

As a famous litterateur in the last years of Han Dynasty, Kong Rong, who is one of the "seven talents in Jian'an (reign title of Emperor Xian)" years, is not remembered for his masterpieces today, but a small story that can even be told by a three-year-old child: Kong Rong gives up the biggest pear.

Pears had already appeared on the dining table as an ordinary fruit in the East Han Dynasty nearly 2,000 years ago. China, as one of the birthplaces, has planted pear trees for more than 4,000 years according to relevant records. In the long-term development process, people continue to explore, improve and cultivate pear varieties. There are records of pears in *Qi Min Yao Shu* (a comprehensive agronomic book), written in the last years of Northern Wei Dynasty, which described the cultivation and grafting techniques and storage methods of pears in detail, showing its high popularization in that period. *Records of Flowers in Luoyang* written in the Song Dynasty recorded 27 varieties of peer trees.

Today, pear trees are planted throughout Henan, Hebei, Shandong, Liaoning Provinces and many other places and rank second after apple trees as the most common fruit in cultivation area and yield. Anshan City, Liaoning, a city though famous for its steel industry, has a place occupied by the Nanguo Pear.

Because Nanguo pear is a unique variety in Anshan area, so it is also called "An Guo", and ranks as one of the "Four Great Pears" in China. The Nanguo pear cultivation system is characterized by planting ancient Nanguo pear trees.

Anshan City is located on the edge of the Northeast China Plain, with the vast flatland in the northwest of the city and mountains accounting for half of the overall area in the northeast. The Monsoon Climate of Medium Latitudes brings distinctive seasons, sufficient rains, heats and sunlight. There are more than 40 rivers and streams, creating a favorable irrigation condition. The ideal growth environment gives the Nanguo pears thin skins, thick and juicy pulp, full fragrance. What is even more rear to see is that the mother plant of Nanguo pears has survived to this day.

On the hillside of Duizhuangshi Village in Dagushan Township, Qianshan District, Anshan City, a short vine-stacked pear tree is vary eye-catching. One side of the tree has rotted, while the other side grows with branches and is decorated with red strips of cloth. This is the mother plant of Nanguo pear trees, the only surviving tree growing from seedling under natural crossing condition, which has been in existence for more than 150 years.

The red strips of cloth on the mother tree of Nanguo pear represent the expectations for a good weather and good harvest next year. Nanguo pear trees do not disappoint the orchard farmers.

南果梨是鞍山地区特有的水果产品。据《中国果树志》（第三卷）记载，现南果梨树母株仍生长于今鞍山市千山区大孤山镇对桩石村，被认定为南果梨祖树，是当地目前已知仅存的一株自然杂交实生苗南果梨树。图为梨祖园。（覃少明 摄）

Nanguo Pear is the specialty in Anshan City. According to the third volume of the *Records of Fruit Trees in China*, the mother tree in Duizhuangshi Village, Dagushan Township, Qianshan District, Anshan City is identified as the "ancestor" of Nanguo pear trees, and the only surviving pear tree growing from seedling under natural crossing condition. The picture shows the ancient orchard of pear trees. Photographed by Qin Shaoming

南国梨的田间养护主要涉及种苗培育、施肥、花果管理和果实采收等环节。物候期为：花芽萌动期4月上旬，初花期4月下旬，盛花期4月末至5月初，落花期5月上旬，落果期5月中旬至6月上旬。图为当地果农在花果管理期给梨树授粉和林下耕作。（覃少明 摄）

The field conservation of Nanguo pear mainly includes germchit cultivation, fertilization, flower and fruit management and fruit harvesting and other steps. The phenological phases are as follows: the buds stage is in early April; the initial bloom stage is in late April; the full-bloom stage is from the end of April to the beginning of May; the falling stage is in early May; the fruiting stage is from the middle May to early June. The pictures show the locals are pollinating and planting in the pear orchard during the flower and fruit management period. Photographed by Qin Shaoming

依靠自身独特的地理、气候条件和栽培经验,鞍山南果梨皮薄肉厚、果肉细腻多汁、香味浓郁,是中国"四大名梨"之一,被誉为"梨中皇后"。图为盛放的梨花和成熟的梨果。(覃少明 摄)

With the favorable geographical, climatic conditions and cultivation experience, the skins are thin, the pulp is thick and juicy and the fragrance is strong, so Nanguo pears are well known for one of the "Four Great Pears" and regarded as the "Pear Queen". The pictures show the blossoming pear flowers and the ripe pears. Photographed by Qin Shaoming

南果梨从起源衍生、人工种植，到现在形成产业链条，每一次的发展和飞跃都与文化内涵息息相关。"南果梨祖树文化"派生出的祈福文化、旅游文化、亲情文化以及文学作品，是延续传统文化的重要载体。挂红布条祈福，与僧道子弟在梨园中论道，已成为南国梨园的一道风景。（覃少明 摄）

Each development and leap of Nanguo pears are closely related to its cultural connotation, from the very first planting, to artificial cultivation, and now an industrial chain is formed. The blessing culture, tourism culture, family culture and literary works derive from the "Ancient Nanguo Pear Tree Culture". It is an important carrier for the continuation of traditional culture. Hanging red strips of cloth for blessing and discussing ideas with monks and priests have become a unique scenery in the pear orchard. Photographed by Qin Shaoming

目前千山区政府制定了相关的保护工作意见，使鞍山南果梨栽培系统这一具有丰富生物多样性和文化多样性，生产与生态功能突出，体现人与自然和谐发展的重要农业文化遗产焕发出新的生机。图为漫山遍野的南国梨以及在梨园中举办中式传统婚礼的新人。（覃少明 摄）

Now, relevant protection plans have been formulated by government of Qianshan District, which makes the Anshan Nanguo Pear Cultivation System, an important agricultural heritage system with rich biodiversity and cultural diversity, outstanding production and ecological functions, harmonious development between human race and nature, now display new vitality. The pictures show that mountains and plains are covered by Nanguo pears; a traditional Chinese wedding of a new couple is held in the pear orchard. Photographed by Qin Shaoming

东北地区 · Northeast China/辽宁省 · Liaoning Province

传统林参共作模式
辽宁宽甸柱参传统栽培体系

在标志着"现代生物时代来临"的第三纪，人参就开始出现并繁衍，是当今的活化石植物之一。中国是最早应用人参的国家。现存最早的中医学专著、成书于东汉时期的《神农本草经》中，已有对人参入药的详细记载。

中医认为，人参具有大补元气、复脉固脱、补脾益肺等功效。在中国，这种外观似人体形状、可入药的植物，被视为"百草之王"。同时，中国也是最早用文字记录人参的地方。距今3500年的殷商时期，"参"字就被刻在了龟甲片上。这个象形文字，与人参的外观特征十分贴合。

人参性喜阴凉，生长在北纬33°—48°之间。在我国东北地区，人参被誉为"东北三宝"之一。它广泛分布于东北地区的东南部至东北部的山区和半山区。辽宁省宽甸满族自治县是人参在东北的产区"南极点"，这片紧邻国境线的区域，是柱参的家乡。

柱参源于野山参，栽培历史久远。据《宽甸县地方志》记载，明万历年间（1610年前后），山东七翁到此采挖野山参，大参拿走，幼参及参籽就地栽种，并栽榆树、立一条石柱为记，一石奠基业，一榆扬旗帜，柱参就此而得名。宽甸柱参传统栽培体系是有400多年历史的半人工栽培柱参系统，以宽甸满族自治县振江镇石柱子村为中心。这里山连绵、水纵横、森林茂密、特产丰富、风景优美，被誉为鸭绿江边的"香格里拉"和"神仙居住的地方"。

宽甸满族自治县有着适宜种植人参的自然条件。这里地处辽宁省南部地区、长白山余脉，林地面积多，森林覆盖率达到78%。温润的季风气候，每年给这里带来充沛的降水。当地参农模拟野山参的生长环境养参。他们培育出的柱参生长到一定年数后，药用价值接近现已濒危的野山参。

近年来，利用当地丰富的森林资源进行林下种植，让柱参的生长回归到自然中去，成为参农首选的种植方式。这一传统的种植方式，达到了参林双赢的效果。这种方式为人参的生长创造了原始的、天然的环境条件，对森林群落的恢复和资源的保护具有深远意义。

地理位置：北纬40°43′—40°50′、东经125°21′—125°28′，地处辽宁省丹东市宽甸满族自治县。
气候特点：温带半湿润季风气候。
认定时间：2013年被农业部列为首批中国重要农业文化遗产。

Geographical location: 40°43′-40°50′N, 125°21′-125°28′E; located in Kuandian Manchu Autonomous County, Dandong City, Liaoning Province.
Climate type: temperate sub-rainy monsoon climate.
Time of identification: In 2013, it was listed as one of the first batch of the China-NIAHS by the Ministry of Agriculture.

The Traditional Forest-ginseng Integrated Model
Kuandian Pillar Ginseng Cultivation System

Ginseng began to appear in the tertiary period which marks the advent of the modern biological era, and now, it is one of the living fossil plants. China is the first country to apply ginseng. The earliest Chinese medicine monographs *Shen Nong's Herbal Classics* written in the East Han Dynasty had already recorded the ginseng medicine in detail.

Herbalist doctors believe that ginseng has the effects of replenishing vital energy, retaining strong pulse and tonifying the spleen and lungs. Therefore, ginseng is regarded as the "Herb King". China was also the first county to record ginseng in words. In the Shang Dynasty about 3,500 years ago, the Chinese character of "参", which was very close to the appearance of ginseng, was engraved on the tortoise shells.

Ginseng grows in shady and cool places, from 33° to 48° north latitude. It is well known as one of the "three treasures" in Northeast China, and widely distributed from the mountains in the southeast of Northeast China to semi-mountainous areas in the northeast of Northeast China. Kuandian Manchu Autonomous County, close to the border line, is the "Antarctic Point" among all the regions planting ginseng. It is also the hometown of pillar ginseng.

The pillar ginseng is derived from wild ginseng and has a long history of cultivation. According to the *Local Chronicles of Kuandian*, during the Wanli years in the Ming Dynasty (about 1610), seven elders from Shandong Province dug ginseng here. They took away the bigger ginsengs and planted smaller ones and seeds in Kuandian. To mark the paces, they planted an elm and set up a stone pillar. The pillar ginseng got its name since then. The traditional Cultivation System of Pillar Ginseng in Kuandian is a semi-artificially cultivated pillar ginseng system with a history of more than 400 years, centered on Shizhuzi Village, Zhenjiang Township, Kuandian Manchu Autonomous County. Scrubland hills, brooky rivers, dense forests, various specialties, graceful landscapes, Kuandian is given the reputation of "Shangri-La" nearby the Yalu River and a superexcellent place god would choose to live in.

Kuandian Manchu Autonomous County has natural conditions suitable for ginseng to grow. It is located in the southern part of Liaoning Province and Changbai Mountains extension, and the forest coverage can reach to 78%. The warm monsoon climate brings plenty of rainfall every year. The local ginseng farms simulate the growth environment of wild ginseng. Then after growing for a certain number of years, the ginseng they cultivated has a medicinal value close to the wild ginseng that is now endangered.

In recent years, planting ginseng under the forest and allowing the ginseng to return to nature have become the preferred planting method for local farmers and achieved win-win effect both for the ginseng and the forest. This method not only creates primitive and natural environmental conditions for the growth of ginseng, but also has profound significance for the restoration of forest communities and the protection of resources.

柱参，亦称石柱人参，系以辽宁省宽甸满族自治县振江镇石柱子村为核心的周边固定区域所独产。石柱子村位于辽宁东部山区鸭绿江畔，与朝鲜隔江相望。图为标示着传统柱参种植的石柱亭和石柱子。（王娟 摄）

The pillar ginseng, also known as stone column ginseng, is a special local product growing only in some certain areas around Shizhuzi Village, Zhenjiang Township, Kuandian Manchu Autonomous County. Shizhuzi Village is located on the banks of the Yalu River in the eastern mountainous area of Liaoning, facing North Korea across the river. The picture shows a stone pavilion and a stone pillar. Photographed by Wang Juan

宽甸柱参源自山东七翁，随着采参的人越来越多，有些人安家在此，逐渐形成了石柱子村。至今，每年的农历三月十六，参农都要立庙祭拜最早养柱参的祖师爷，为其过生日。图为石柱子和参农祭拜养柱参祖师爷的场景。（王娟 摄）

With the spread of the story of the seven elders from Shandong Province, more and more people began picking ginseng or settled here, and gradually formed Shizhuzi Village. Up to now, every year on the 16th day of the lunar month March, the locals will worship the ancestors who first raised pillar ginseng to celebrate their birthdays in a temple. The pictures show the scene of a stone pillar and locals worshipping the ancestors. Photographed by Wang Juan

近年来，林下种植成为首选的种植方式，柱参生长得以回归自然，不仅参林双赢，而且资源得到永续利用。柱参传统栽培技艺已列为辽宁省非物质文化遗产。图为今天的柱参种植和当地参农察看当年参苗成长情况。（王娟 摄）

In recent years, planting under the forest has become the preferred method, which allows the ginseng to return to nature. Not only the forest and the ginseng can grow together, but also the resources can be kept sustainable. The traditional cultivation techniques of ginseng have been listed as an Intangible Cultural Heritage List of Liaoning Province. The pictures show that today's ginseng planting and the farmers are checking the ginseng condition. Photographed by Wang Juan

柱参的参籽也是名贵的药材，而且关乎柱参的繁殖。参籽初生时为绿色，成熟后变为浆果红色。（王娟 摄）

The seeds of the ginseng are also valuable medicinal materials, and concern the reproduction of the ginseng. The seeds are green when they are born, and become berry-like red when they are ripe. Photographed by Wang Juan

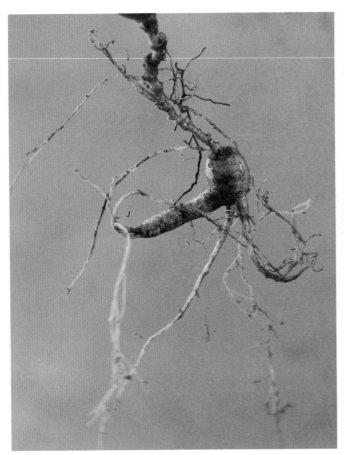
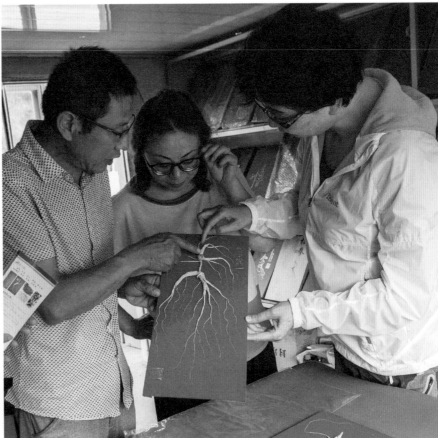

柱参芦高体灵、皮老纹深、须长须清、珍珠疙瘩多、形态优美。400多年来，经历代参农培育，已形成圆膀圆芦、草芦、线芦、竹节芦4个特有品系，成为人参家族的一个独特种类。图为石柱参和当地商户向游客推销石柱参。（王娟 摄）

The pillar ginseng is in a beautiful shape: long ginseng body, plicate skin, long and clear beard, pearled-shaped pimples. For more than 400 years with continuous breeding from generation to generation, there have been four varieties of the pillar ginseng: round ginseng, grass ginseng, string ginseng, bamboo-joint ginseng. Obviously, the pillar ginseng is a unique variety among the ginseng family. The pictures show that the local merchants are selling ginseng to tourists. Photographed by Wang Juan

近年来，宽甸加大了对域内环境的保护力度，优厚的自然环境有助于柱参产量和品质的提升，其中石柱子村片区水土保持工程为柱参的持续发展提供了保证。（王娟 摄）

In recent years, Kuandian has strengthened the protection of the environment within the region, and the favorable natural environment is conducive to the improvement of the yield and quality of pillar ginseng, among which the soil and water conservation project in the area of Shizhu Village provides a guarantee for the sustainable development of pillar ginseng. Photographed by Wang Juan

华东地区·East China

江苏省

沼泽洼地土地利用模式——江苏兴化垛田传统农业系统

浙江省

传统稻鱼共生农业生产模式——浙江青田稻鱼共生系统

陡坡山地高效农林生产体系——浙江绍兴会稽山古香榧群

传统茶禅文化代表——浙江杭州西湖龙井茶文化系统

低洼地区杰出的生态农业模式——浙江湖州桑基鱼塘系统

世界香菇起源地——浙江庆元香菇文化系统

福建省

湿地山地立体农业生产体系——福建福州茉莉花与茶文化系统

竹林、村庄、梯田、水系综合利用模式——福建尤溪联合梯田

乌龙茶发源地——福建安溪铁观音茶文化系统

江西省

世界人工栽培稻源头——江西万年稻作文化系统

最大的客家梯田——江西崇义客家梯田系统

山东省

沙地生态治理与经济发展的典范——山东夏津黄河故道古桑树群

沼泽洼地土地利用模式
江苏兴化垛田传统农业系统

在农业生产中，除了向山要地（梯田），人们还与水争地、向水要田，垛田便是生活在江苏兴化里下河地区的人们向水要来的田地。

垛田是一种由泥土堆积而成、高出水面的台状高地，是兴化地区独有的土地利用方式。大大小小的垛田田块如同海洋中的岛礁一般，遍布在兴化的沼泽湿地湖荡之中。

兴化地处江苏中部，河网纵横、湖泊密布。与周围地区相比，兴化地势低洼，如同"锅底"。每到汛期，兴化周围的"四湖"（洪泽湖、高宝湖、白马湖、邵伯湖）、"三河"（里运河、通扬运河与淮河）、"一海"（黄海）的水便一齐向兴化涌来。洪水一来，兴化地区的田地被冲、庄稼被毁，农民深受其害。为了减少洪水带来的损失，宋元时期起，兴化人开始修筑垛田，在垛田上进行农业种植。垛田的雏形是架田——在沼泽地用木桩、木架塞上泥土水草，覆盖土壤，形成田块。后来，人们选择沼泽湿地中的地势稍高处，用泥土堆积起来，形成高出水面1米以上的田块。垛田的形状因河沟宽窄而变，大的不过数亩，小的仅有几分。

在还没有化肥的年代，在垛田上劳作的农民，把疏浚河沟挖出来的泥浆和杂草，作为有机肥堆积到垛田上（这种方式被称作"罱泥""扒苲"），使得垛田以每年几厘米的速度逐渐增高。高出水面数米的田块，在洪水来临时，丝毫不受影响，为兴化的农业生产提供了庇护。同时，垛田的地势高，排水良好。罱泥、扒苲的堆肥方式，使得垛田的土壤疏松，土质肥沃，各种作物在垛田上都能长势良好。

近代以来，随着对洪涝灾害治理手段的改进，兴化地区大面积受洪水侵害的概率大幅变小。同时，社会稳定，人口增加，数米高的垛田被农民削矮，向周围连片扩展。如今兴化垛田总面积已达47万亩，核心区6万亩，被农民种上了大面积的油菜。每当油菜花开的季节，湖荡里的块块垛田全是黄灿灿的一片，连成了"中国最美油菜花海"。2009年起，兴化每年3至5月菜花飘香时节都会举办菜花文化节，吸引了众多中外游客参观游玩。

地理位置：北纬32°40′—33°13′、东经119°43′—120°16′，地处江苏省泰州市兴化市。
气候特点：亚热带湿润性季风气候。
认定时间：2013年被农业部列为首批中国重要农业文化遗产，2014年被联合国粮农组织列为全球重要农业文化遗产。

Geographical location: 32°40′-33°13′N, 119°43′- 120°16′E; located in Xinghua City, Taizhou City, Jiangsu Province.
Climate type: subtropical humid monsoon climate.
Time of identification: In 2013, it was listed as one of the first batch of the China-NIAHS by the Ministry of Agriculture. In 2014, it was listed as one of the GIAHS by the FAO.

The Land Use Patterns in Swamp and Lowland Areas
Xinghua Duotian Traditional Agrosystem

In agricultural production, people have to compete with mountains or waters for fields or terraces. Duotian is the war trophy of the people living in Lixiahe area, Xinghua City have won.

Duotian, platform-like highlands risen above the water by mud, is a unique land use method in Xinghua. Duotian in all sizes scatter among the marches and wet lands like rocks above the sea level.

Xinghua is located in the central part of Jiangsu Province and has a wide network of rivers and lakes. Compared with the surrounding area, Xinghua is in a lower terrain, like the bottom part of a pan. Every time in flood seasons, Xinghua will witness an influx of waters from the "Four Lakes" (Hongze Lake, Gaobao Lake, Baima Lake, Shaobo Lake), the "Three Rivers" (Li Canal, Tong-Yang Canal, Huai River) and the "One Sea" (Yellow Sea). When the flood came, farmers in Xinghua suffered a lot since the fields were rushed and the crops were destroyed. In order to reduce the losses caused by the floods, since the Song and Yuan Dynasties, Xinghua people began to build Duotian and planted crops on it. The rudiment of Duotian was rack field—fields were divided into many bars by racks with mud and aquatic plants covering the underneath soil. Later, people chose the higher places in the marshes and wetlands and piled up with mud to form lands that were more than 1 meter above the water surface. The size varies depending on the width of the river ditches, some of the Duotian are a few acres while the smaller ones can be only one-tenth of the bigger ones.

In the age when there was no fertilizer, the farmers would dumped the mud and weeds dug out from river ditches as organic fertilizer on the Duotian, making it gradually rise at a speed of several centimeters per year. The lands meters above water surface would not be influenced at all when the flood comes. Therefore, the agricultural production in Xinghua was well protected. High terrain gives Duotian a good drainage. The mud and weeds fertilizer make the soil loose, fertile and suitable for all kinds of dryland crops to grow well.

In modern times, with the improvement of flood regulation measures, the probability of large-scale damage in Xinghua area has been greatly reduced. At the same time, Duotian several meters high are shortened while expanding to the surrounding areas to feed the increasing population. Now the vast Duotian in Xinghua, with 470,000 mu of its total area including 60,000 mu of the core area, is covered by a wide range of rape. Each year in the flower season, all the yellow bars patch together into "The Most Beautiful Sea of Rape Flower in China". Since 2009, each year in March and May when the flowers bloom, a culture festival of rape flowers will be held, which attracts many Chinese and foreign tourists to visit.

华东地区・East China/ 江苏省・Jiangsu Province

江苏兴化垛田传统农业系统，是农业、林业、渔业相结合，在独具特色的沼泽洼地中巧妙利用垛形土地的湿地生态农业系统，已有上千年的造田耕作历史。（奚鸣君 摄）

The Xinghua Duotian Traditional Agrosystem is a wetland ecological agricultural system featuring the combination of agriculture, forestry and fishery. It utilizes swamps and wetlands to build stack-shaped fields, which has a history of thousands of years of reclaiming cultivation. Photographed by Xi Mingjun

兴化垛田因湖荡沼泽而生，每块面积不大，形态各异，大小不等，四周环水，各不相连，形同海上小岛，人称"千岛之乡"。核心区有6万多亩这样的耕地，分布在垛田镇、缸顾乡、李中镇、西郊镇、周奋乡一带。（朱明 摄）

Duotian are built based on the terrains, so each pieces of land is small in size and different in shape. They are surrounded by water and not connected. "Hometown of Thousands of Islands" is an alternative name of Xinghua because the pieces of lands are like countless islands above the sea level. There are more than 60,000 mu of this kind of fields, which are mainly distributed in Duotian, Ganggu, Lizhong, Xijiao and Zhoufen Townships. Photographed by Ju Ming

垛田是兴化水乡特色之一。每年春季，油菜花开，蓝天、碧水、"金岛"织就了"河有万弯多碧水，田无一垛不黄花"的奇丽画面。泛舟其中，如入迷宫，浓郁花香让人迷醉，旖旎风光令人流连忘返。（上图 万亮 摄；下图 杨秋红 摄）

Duotian is one of the characteristics of the water town Xinghua. The rape flowers, the blue sky, the clear river, the "golden island", "there is no single stream without thousands of bends, there is no single field without golden yellow flowers", as one poem described the magnificent landscape. Drifting in the fields likes entering into a maze, sniffing the fragrance likes getting a drunk and unwilling to leave.The picture above photographed by Wan Liang, and below by Yang Qiuhong.

至今，垛田还保存着传统的农耕方式，用天然生态肥料种植蔬菜。农人的田间劳作都要泛舟而行，家家有船，户户荡桨。图为当地农民乘坐木船浇灌垛田上的作物。（吴萍 摄）

The traditional farming method by using environmentally friendly fertilizer is preserved until now. Dinghies are necessity for every household to "walk" in the water field. The pictures show local farmers watering the crops in a wooden boat. Photographed by Wu Ping

兴化结合乡村旅游和摄影人创作需要开发了千岛菜花景区,每年举办千岛菜花节,其中乘船游也是到这里游玩的游客们的必选项目。(杨天民 摄)

To satisfy the need of tourists and photographers, scenic area featuring the rape flowers in the "islands" has been developed. The cultural festival of rape flowers is held every year, and the boat trips are also a must for visitors to the area. Photographed by Yang Tianmin

随着旅游经济的发展，聪明的兴化人民利用垛田这种独特的形式，从事大规模油菜生产，发展休闲观光农业。万岛耸立，千河纵横，可谓天下奇观。连续举办多届的中国兴化千岛菜花旅游节已经成为享誉全国的新兴旅游亮点。（杨天民 摄）

With the development of the tourism economy, the bright Xinghua people have used the unique Duotian landform to engage in large-scale rape production and develop recreational agriculture. With thousands of pieces of lands, interlaced rivers and streams, the scene can be count as a wander all over the world. Xinghua, a city for holding the cultural festival of rape flowers in the thousands of "islands", has been a famous tourism highlight in China. Photographed by Yang Tianmin

华东地区・East China/江苏省・Jiangsu Province

20世纪90年代以来,随着蔬菜加工工业的发展,垛田的田间作物逐步向香葱等加工原料调整转移,油菜籽种植日渐减少,但生物多样性获得发展。图中的作物为当地著名的龙香芋。(班映 摄)

Since the 1990s, with the development of the vegetable processing industry, the field crops planted in Duotian have gradually shifted to the processing materials such as scallions, and the cultivation of rapeseeds has been daily decreased, but the biodiversity has been developed. The crop in the picture is the famous "dragon fragrance taro". Photographed by Ban Ying

垛田独特的岛状耕地，是荒滩草地堆积而成，土质疏松、养分丰富，加上光照足、通风好、易浇灌、易耕作，使得生产的蔬菜无论是品质还是产量，都是普通大田种植难以比拟的。如今夏日里的兴化垛田，油菜、香芋、香葱、韭菜等经济作物构成了当地除旅游资源外的丰富经济来源。（杨天民 摄）

The unique island-like land stacked by mud and weeds has the advantages of loose soil, rich nutrients, ample sunlight, good ventilation, easy irrigation and cultivation, which make the vegetables growing in common field have nothing on the vegetables growing in Duotian, no matter in quality or yield. In addition to scenic spots, some cash crops such as the edible rape, the taro, green Chinese onion, garlic chives become the financial sources for the local farmers. Photographed by Yang Tianmin

传统稻鱼共生农业生产模式
浙江青田稻鱼共生系统

红鲤、草鱼、鲫鱼、田鱼等造型的鱼灯，插在杖棍上，在舞者手中高举着，从两侧鱼贯而出。伴随着音乐的鼓点，舞者们不断变换着阵势：编篱阵、二龙喷水阵、抲三拗、四角循、五梅花、六角循……整支舞热烈而又朴素，在舞者的上下翻飞中，"鱼儿们"如在水中一般灵活和自由，时而摇头摆尾，时而起伏游弋。

这是浙江省丽水市青田县最具代表性的民间舞蹈——青田鱼灯舞。青田鱼灯舞曾参加首都新中国成立50周年庆典、第五届中国国际民间艺术节、第七届中国艺术节和中西建交30周年庆典、北京奥运会、上海世博会、米兰世博会、第八届全国残运会、中意建交40周年庆典等国内外文化交流活动，2008年被列入国家非物质文化遗产名录。

青田鱼灯舞的编排创意来自于青田县独特的农业生产模式——稻鱼共生。当地农民根据水稻的生长周期，把鱼放到稻田里饲养，可以同时收获稻米和田鱼。公元9世纪，青田就出现了这种农业生产方式，它以种养结合为特征，一直延续至今。

青田地处瓯江流域中下游、浙江省东南部的丘陵地区，境内多山少田，有着"九山半水半分田"之说，是传统的水稻种植区。最初，当地的农民引来山上的溪水灌溉稻田。溪中的小鱼苗，顺着水流被引进了稻田中，并在稻田里顺利生长。待鱼长大到脊背露出水面之时，便会被农民发现并捕获。在稻田里长大的鱼，味道鲜美无比。于是，人们开始往稻田里人工投放鱼苗。每年插秧的时节，也是鱼苗被投入稻田的时候。在接下来六七个月的时间里，水稻为田鱼提供食物和栖息场所，田鱼则为水稻松土增肥、消灭害虫。在这个过程中，水稻扮演了生产者的角色，鱼类和其他水生动物是消费者，而溪水中存在的细菌和真菌则是分解者。三者互利互惠、各司其职，构成了一个不需任何外界物质介入的自循环生态系统。

在这种"鱼食昆虫杂草—鱼粪肥田"的模式中，农业生产所需的材料在大自然中都能找到，几乎不需要使用化肥农药，不仅保证了稻田湿地的生态平衡，也解放了当地农民的生产力。

地理位置：北纬27°56′—28°29′、东经119°47′—120°26′，地处浙江省丽水市青田县。
气候特点：亚热带季风气候。
认定时间：2005年被联合国粮农组织列为全球重要农业文化遗产，2013年被农业部列为首批中国重要农业文化遗产。

Geographical location: 27°56′-28°29′N, 119°47′-120°26′E; located in Qingtian County, Lishui City, Zhejiang Province.
Climate type: subtropical monsoon climate.
Time of identification: In 2005, it was listed as one of the GIAHS by the FAO.In 2013, it was listed as one of the first batch of the China-NIAHS by the Ministry of Agriculture.

A Traditional Rice-fish Symbiotic Agricultural Production Model
Qingtian Rice-Fish Culture System

The dancers hold the lanterns, which are plugged in sticks and shaped likes common carp, grass carp, crucian carp, and paddy field fish, snaking from both sides and changing their positions with the music from the "fence array", to "two dragons spraying water", "three bends", "four circles", " five petal of the plum blossom", "six circles"...The dance is impassioned sometimes, calm sometimes; the fishes are like swimming in the water with the dancer's ups and downs, shaking their heads and tails sometimes, undulating sometimes.

This is the most representative folk dance in Qingtian County—fish lantern dance. It was once shown on the celebration of the 50th Anniversary of the Founding of the People's Republic of China in capital Beijing, the 5th International Folk Art Festival, the 7th China art festival and the 30th Anniversary of Establishing Diplomatic Relations between China and Spain, the Beijing Olympic Games, the Shanghai World Expo, the Milan World Expo, the 8th National Paralympic Games, the 40th Anniversary of the Establishment of Diplomatic Relations between China and Italy, and other cultural exchange activities at home and abroad. In 2008, it was listed in the National Intangible Cultural Heritage List.

The originality of the fish lantern dance comes from the unique agricultural production mode of Qingtian County—the rice-fish symbiosis. According to the growth cycle of the rice, local farmers raise fish in rice fields and harvest rice and fish growing in field at the same time. This kind of agricultural production mode appeared in Qingtian in the 9th century. It is characterized by combination of planting and breeding and has continued to this day.

Qingtian is located in the middle and lower reaches of the Ou River and in the hilly areas in the southeast of Zhejiang Province. Covered mainly by mountains and hills, Qingtian is a traditional planting zone of rice. At the very beginning, local farmers diverted streams from the mountains to irrigate rice fields. The little fish in the stream were led into the rice fields along the streams and grew smoothly. When the back was out of the water, those fish would be found and captured by the farmers. The fish growing up in the rice fields were delicious. Hence, people began to artificially put the juvenile fish in the rice fields.

When it comes to the time to transplant rice seedlings, the juvenile fish are also put into the rice fields. In the next six or seven months, the rice provides food and habitat for the paddy field fish, while the paddy field fish fertilize, loosen the soil and help eliminate pests. In this process, rice plays the role of producer, fish and other aquatic animals are consumers, and bacteria and fungi in the stream are decomposers. The three mutually benefit each other and perform their own duties, forming a self-circulating ecosystem that does not require any external material intervention.

In this model of "fish-insect-weed-fish", the materials needed for agricultural production can be found in nature, and there is no need to use chemical fertilizers or pesticides, which not only guarantees the ecological balance of rice paddy wetlands, but also liberates the productivity of local farmers.

青田县位于浙江省东南部，瓯江流域的中下游。该县自公元 9 世纪开始一直保持着传统的农业生产方式——"稻田养鱼"，并不断发展出独具特色的稻鱼文化。图为当地农民利用田鱼除草，施肥，除虫，促进生态平衡。（汤洪文 摄）

Qingtian County is located in the southeastern part of Zhejiang Province, the middle and lower reaches of the Ou River Basin. Since the 9th century, the county has maintained the traditional rice-fish agricultural production way and has continued to develop a unique rice-fish culture. The picture shows that local farmers are using the fish to weed, fertilize and deworm to promote ecological balance. Photographed by Tang Hongwen

龙现村，位于青田县城西南部方山乡境内，距县城 20 千米，背依奇云山，这里是稻鱼共生系统全球重要农业文化遗产保护地。龙现村有 1200 多年的稻田养鱼历史，全村有养鱼梯田 400 多亩，水塘 140 多个，具有得天独厚的养殖田鱼优势。稻田养鱼已成为当地农民祖辈相传的种养习惯。（汤洪文 摄）

Longxian Village is located in Fangshan Township, southwest of Qingtian County. It is 20 kilometers away from the Qingtian County and backed by Qiyun Mountain. It is an GIAHS site for the rice-fish symbiosis system. Longxian Village has a history of more than 1,200 years of breeding fish in rice fields. There are more than 400 mu of fish terraces in the village and more than 140 ponds. With particularly favourable natural conditions, the rice-fish model has been maintained and passed on from generation to generation. Photographed by Tang Hongwen

青田田鱼品种优良，肉质细嫩，鳞软可食，是观赏、鲜食、加工的优良彩鲤品种。龙现村村民的房前屋后、田间地头，凡是有水的地方，不论稻田、水渠、水沟、水池、水塘，也不论水深水浅，都养殖田鱼。（汤洪文 摄）

The paddy field fish in Qingtian has tender meat, soft and edible scales, so it is an excellent variety of koi fish for ornament, fresh-eating and processing. Fish will be raised wherever there is water, no matter in the rice fields, the canals, the ditches, the pools or the ponds, no matter how deep or shallow the water is. Photographed by Tang Hongwen

青田好山，好水，好空气，为田鱼生长提供良好生态环境。一分耕耘，一分收获。近年青田稻田养鱼实现每亩千斤粮，百斤鱼，收入超万元。（汤洪文 摄）

The mountains, the rivers and the air all nurture the growth the paddy field fish. The more you plough, the more you gain. In recent years, the rice-fish culture system in Qingtian can harvest one thousand jin of rice and one hundred jin of fish per mu, with the income of over 10,000 yuan. Photographed by Tang Hongwen

华东地区·East China/浙江省·Zhejiang Province

青田鱼干是青田鱼一种非常重要的加工制品,采用具有千年历史的传统工艺制作烘烤,体溢芳香。农户传统制作的这些田鱼干香脆可口,成为馈赠亲友的珍贵礼品。(汤洪文 摄)

Qingtian dried fish is the local product roasted by a traditional craftsmanship which has been in existence for more than one thousand years. These fish is crispy and delicious, and thus it is a precious gift for friends and relatives. Photographed by Tang Hongwen

青田稻鱼米秉承"青田稻鱼共生系统"的种养模式，按照统一规划、统一品种、统一种植标准、统一加工和统一品牌包装销售的原则，使用区域公共品牌。青田稻鱼米、生态米与青田鱼一起成为地方特产。（青田县农业局　供图）

The products from the rice-fish culture system come into the market in unified variety, planting or breeding benchmark, processing steps, regional public bands. Qingtian rice raised with the fish and the ecological rice together with Qingtian fish become local specialties. Provided by Agriculture Bureau of Qingtian County

悠久的田鱼养殖史还孕育了灿烂的田鱼文化，青田田鱼与青田民间艺术结合，派生出了一种独特的民间舞蹈——青田鱼灯舞，以及其他民俗文化活动。图为每年秋收季节青田农村举行祭祀活动表达丰收的喜悦。（汤洪文 摄）

The long history of fish farming has also bred a splendid fish culture. The combination of fish and folk art has given birth to a unique folk dance—the fish lantern dance, and other folk cultural activities. The picture shows that each year in the autumn harvest season, sacrificial activities are held in the countryside to celebrate the good harvest. Photographed by Tang Hongwen

陡坡山地高效农林生产体系
浙江绍兴会稽山古香榧群

苏轼在杭州任职时，为他的朋友郑户曹写了一首诗：《送郑户曹赋席上果得榧子》。诗中写道："彼美玉山果，粲为金盘食。瘴雾脱蛮溪，清樽奉佳客……"他希望郑户曹像玉山果一样，德行美好如油脂般可以滋润自己。这被苏轼大为认同和赞赏的玉山果，指的就是浙江绍兴一带的特产——香榧树的果实香榧子。

香榧树是由人工嫁接榧树而来。榧树有着一个成员遍布世界多地的大家族，其树干作为良材在木材市场上广受欢迎。香榧则是唯一一种果实可食用的榧树，也是我国独有的榧树品种。

想获得香榧树的果实香榧子确实需要一些耐心。时间在香榧面前仿佛静止不前：生长成熟期极为漫长，第一年开花，第二年结果，第三年成熟后方可得香榧子。然而，在香榧树开花之前，还需无尽的等待。

绍兴会稽山脉，海拔高，降水多，对于爱光而又喜好凉爽湿润的榧树来说，是极佳的生长地。据传，西汉时期，居住在今绍兴会稽山附近的山民，发现有两棵野生榧树纠缠着长到了一起，而这棵"合成树"结出的果实竟然可以食用。于是，居住在会稽山附近的人们，开始对野生榧树进行嫁接培育。现存古香榧树基部多有显著的"牛腿"状嫁接疤痕。古香榧树历经千年仍硕果累累，堪称古代良种选育和嫁接技术的"活标本"。不过，根据资料和现存最古老的香榧树的树龄推测，对香榧树的人工栽培大概起于唐代，在明清时期开始大规模发展。

到现在，会稽山脉的香榧群已达400平方千米。香榧群中有结实香榧大树10.5万株，百年以上的古香榧树超过7.2万株，千年以上的古香榧树有数千株，被称为"榧树王"的香榧树树龄已达1430年。这些古香榧树的基部都有明显的"牛腿"状嫁接疤痕，给现代人们研究古代良种选育和嫁接技术提供了活标本。在长期的发展中，当地居民逐渐形成了"香榧树—梯田—林下作物"的复合经营体系。他们在山地上构筑梯田，种植香榧树，在香榧林间种植茶叶、杂粮等作物，构成了独特的水土保持和高效产出的陡坡山地利用系统。

地理位置：北纬29°42′—30°19′、东经120°16′—120°46′，地处浙江省绍兴市。
气候特点：亚热带季风气候。
认定时间：2013年被联合国粮农组织列为全球重要农业文化遗产，同年被农业部列为首批中国重要农业文化遗产。

Geographical location: 29°42′-30°19′N, 120°16′-120°46′E; located in Shaoxing City, Zhejiang Province.
Climate type: subtropical monsoon climate.
Time of identification: In 2013, it was listed as one of the GIAHS by the FAO. In the same year, it was listed as one of the first batch of the China-NIAHS by the Ministry of Agriculture.

The High-efficiency Agroforestry Production System Applied for Slopes
Shaoxing Kuaijishan Ancient Chinese Torreya

When Su Shi was during his tenure in Hangzhou, he wrote a poem for his friend Zheng Hucao. He wished that the virtues of Zheng Hucao could moisten himself like the oil moisten others. The Yushan fruit, which was greatly recognized and appreciated by Su Shi, refers to the specialty of Zhejiang—the fruit of Chinese torreya tree.

The Chinese torreya tree is attained by grafting. Torreya trees are widely planted around the world, and their trunks enjoy a goof timber market. The Chinese torreya is the only torreya whose fruits are edible, and it is also a unique torreya variety in China.

It takes fairly long time to harvest the fruit of the Chinese torreya. Time seems to be frozen in front of them: there is endless wait before the blooms (first year), stats fruiting (second year) and harvesting (third year).

With sufficient rainfall and in a high altitude, Kuaiji Mountain is an excellent habitat for Chinese torreya since they prefer cool and moist environment with abundant sunlight. It is told that during the Western Han Dynasty, the people living near the Kuaiji Mountain in Shaoxing found two wild Chinese torreya entangled together. Surprisingly, fruits from the "composite tree" were actually edible. Since then, people began to graft and cultivate Chinese torreya. There are many showy "bull leg-shaped" scars caused by the graft at the base of the existing ancient Chinese torreya. The two are the best example of how to keep fruitful for thousands of years after ancient breeding and grafting. According to some data and the age of the oldest Chinese torreya, it is estimated that the artificial cultivation could date back to the Tang Dynasty and began to develop on a large scale during the Ming and Qing Dynasties.

Up to now, there is about 400 square kilometers of Chinese torreya in the Kuaiji Mountain, including 105,000 grown trees, more than 72,000 trees over a hundred years old and thousands of ancient trees over thousands of years old. The oldest Chinese torreya known as "Torreya King" is 1,430 years old. These showy bull leg-shaped scars provide "living specimens" for scholars to study ancient breeding and grafting techniques.

In the long-term development, the local residents gradually form a compound management system of "Chinese torreya-terraces-crops". They build terraces and plant Chinese torreya on the mountains, plant tea bushes and crops in torreya forests, which structures a unique slope-mountain utilization system with efficient water and soil conservation and high yield.

绍兴会稽山古香榧群位于绍兴市域中南部的会稽山脉（主脉在绍兴市的诸暨市、嵊州市、柯桥区、越城区以及金华市的东阳市，平均海拔 500 米）。图为诸暨市赵家镇榧王村的"中国香榧王"，树龄 1380 多年，树高 20 米，胸围 9.26 米，覆盖面积 500 多平方米，历经千年仍硕果累累。（郦以念 摄）

Shaoxing Kuaijishan Ancient Chinese Torreya is located in Kuaiji Mountain and southern part of Shaoxing City (the main range is in Zhuji City, Shengzhou City, Keqiao District, Yuecheng District, Dongyang City in Jinhua Prefecture-level City, with an average elevation of 500 meters). The picture shows the "Torreya King" in Feiwang Village, Zhaojia Township, Zhuji City. It is 1,380 years old, 20 meters high, 9.26 meters of its girth, and covers 500 square meters. It is still fruitful even after one thousand years. Photographed by Li Yinian

香榧四季常绿、形态优美，一棵棵古香榧树，与古村落、小溪、山岚等构成了一幅幅令人赏心悦目的画图。这珍贵的遗产吸引了众多中外游客参观游览，居住其间的人们尽享自然和谐之美。（绍兴市会稽山古香榧群保护管理局 供图）

The evergreen Chinese torreya in graceful shape, the antique village, running clear rivers and rolling hills, combined with the cultural heritage all feast the eyes of every tourist from all around the world. Provided by Conservation bureau of ancient torreya torreya in kuaijishan, shaoxing city

香榧树能够千年不衰，得益于会稽山自古至今的优良生态系统。绍兴先民利用陡坡山地，构筑梯田（鱼鳞坑），种植香榧树，香榧林下间作茶叶、杂粮、蔬菜等作物，构成了独特的水土保持和高效产出的陡坡山地利用系统。因海拔高度不同，会稽山古香榧群的分布呈现聚散结合的特点，既有满山成片分布的，也有离散分布于村落房屋附近的。（宋丹丹 摄）

Thanks to the excellent ecosystem of Kuaiji Mountain since ancient times, the Chinese torreya can survive and flourish to this day. Shaoxing ancestors used the steep slopes to build terraces (known as fish-scale pits), plant the Chinese torreya, and intercrop with tea bushes, crops and vegetables, which then forms a unique mountain-slop utilization system with efficient water and soil conservation and high yield. Based on different altitudes, the Chinses torreya is planted around a certain area or scattered near the village. Photographed by Song Dandan

香榧属裸子植物，无真正的果实，其种子呈核果状，习惯称果实，外被肉质假种皮包裹，绿色，熟时淡黄色至暗紫色或紫褐色，外有白粉。图为生长中的香榧和采摘后并经过腐烂去皮的香榧果。（上图 绍兴市会稽山古香榧群保护管理局 供图；下两图 郭斌 摄）

The Chinese torreya belongs to gymnospermae. It has no real fruit, but has drupaceous seeds, which are commonly known as its fruit, covered by fleshy green aril. The seed changes from light yellow to dark purple or purple brown with white powder outside during its maturation stage. The pictures show the growing Chinese torreya and its fruit which has been rotted and peeled after picking.The picture above provided by Conservation bureau of ancient torreya torreya in kuaijishan, shaoxing city, and below by Guo Bin

香榧有一个有趣的名字——三代果。香榧一树长三果，一年收一次，但同时明年的香榧已经长上枝头了，就是说今年开花结果明年成熟，明年开花结果后年成熟。香榧采摘全靠人工，当地榧农使用云梯（当地称"蜈蚣梯"）等工具采收当年成熟的香榧。（左图 郭斌 摄 右图 绍兴市会稽山古香榧群保护管理局 供图）

People name the fruit of Chinese torreya "Sandaiguo", which means fruits of three generations on one tree. The ripe fruit will be picked once each year, but at the same time, some youth fruit that should be harvested next year has been growing on the tree already. That is to say, bloom and start to fruit this year, then pick the fruit next year. The picking work is all down by hand. The locals use tools such as the ladder (locally called "chilopod ladder" for its length) to pick the ripe fruit. The picture left photographed by Guo Bin and right provided by Conservation bureau of ancient torreya torreya in kuaijishan, shaoxing city

香榧采收后需要待青皮腐烂后去皮成紫褐色，再经清洗晾晒炒制加工。自古至今，赞美香榧的散文、诗歌、美术作品层出不穷，相关祭祀、节庆等活动丰富多彩。图为近年绍兴各地举办的香榧文化节、香榧民间祭祀活动、香榧炒制大赛场景以及 2016 年底开馆的中国香榧博物馆。（绍兴市会稽山古香榧群保护管理局　供图）

The fruit needs to be peeled after harvest, and then it will tend into purple brown; next, it will be cleaned, dried and stir-fried. Endless essays, poems, works of art, and sacrificial activities or festivals are related to praises of the Chinse torreya. The pictures show some folk sacrificial activities held from place to place, stir-frying competition and the Chinese Torreya Museum opened at the end of 2016. Provided by Conservation bureau of ancient torreya torreya in kuaijishan, shaoxing city

传统茶禅文化代表
浙江杭州西湖龙井茶文化系统

这世界上大概没有一个国家的人像中国人这么喜爱茶。这种低矮的绿色植物的叶子，在中国人的生活中，是一种稀松平常而又不可或缺的存在。人们饮茶，由此延伸出制茶、品茶、用茶、赏茶、诵茶、为茶制造各种器皿等一系列茶文化。但是，中国人到底是什么时候开始把茶叶当作饮品的？现在众说纷纭，已不可考。中国现存历史最久、最完备介绍茶的一部书，是唐代时陆羽写作的《茶经》。这本书翔实地记载了前人和当时人的茶叶生产经验，从茶叶的起源，讲到了饮茶的器皿和品茶法。可见，陆羽写《茶经》之时，中国的茶文化已经趋近成熟。《茶经》中的第八个部分，记载了当时中国的茶叶产区。在介绍浙西产区的文字中，有这样一句话："钱塘生天竺、灵隐二寺。"这是说今杭州的天竺寺、灵隐寺是钱塘地区的茶叶产区。这是目前已知最早的对杭州产茶的记载。

北纬30°是世界公认的茶叶产区黄金带。杭州西湖，恰好被这条黄金带穿过，是龙井茶的故乡。杭州龙井茶文化系统就是以龙井茶品种选育、种植栽培、植保管理、采制工艺和茶文化为核心的农业生产系统，以及该系统在生产过程中孕育的生物多样性、发挥的生态系统功能、呈现的人文和自然景观特征。

西湖西南畔的群山上空，常常氤氲着从西湖飘来的水汽。连绵的群山阻挡了北来的寒气和南来的暖流，氤氲的水汽在炎热的亚热带气候中，既遮挡了部分光照，又提供了足够的湿度。龙井茶树便在这独特的地理小气候中生长，以此培育了龙井茶叶色翠、形美、香郁、味醇的特点，以及"淡而远、香而清"的独特味道。"西湖龙井茶"之名始于宋，闻于元，扬于明，盛于清。到今天，西湖龙井茶已经冠列中国名茶之首。除了得天独厚的地理条件外，考究的采制技术也是成就西湖龙井茶的重要原因。西湖龙井茶的采摘有三大特点：早、嫩、勤。鲜叶嫩、匀、绿是龙井茶的基本要求。在鲜叶采摘回来后，还要经过八道工序才能制成成茶。炒制过程中，对"抓、抖、（透）搭、拓（抹）、捺、推、扣、甩、磨、压"十大炒制手法的灵活运用，是考验一个茶农是否成熟的标志。在多年的发展中，当地的茶农对茶叶的细采精炒有着一整套成熟的生产技术，他们世代守着自己的一片茶园，种植栽培、采摘炮制的手艺和精致的茶文化在他们中间代代相传。

地理位置：北纬30°04′—30°20′、东经119°59′—120°09′，地处浙江省杭州市。

气候特点：亚热带季风性湿润气候。

认定时间：2014年被农业部列为第二批中国重要农业文化遗产。

Geographical location: 30°04′-30°20′N, 119°59′-120°09′E; located in Hangzhou City, Zhejiang Province.
Climate type: subtropical humid monsoon climate.
Time of identification: In 2014, it was listed as one of the second batch of the China-NIAHS by the Ministry of Agriculture.

A Representative of Traditional Tea-zen Culture
Hangzhou West Lake Longjing Tea Culture System

It is hard to find another county who is infatuated with tea like China. The little leaves growing on short green bushes are a common while indispensable necessity in the lives of Chinese people. A series of tea cultures such as tea-making, tea-tasting, tea-appreciating, tea-reciting and tea set all derive from tea-drinking. However, it is not clear when the Chinese begin to drink tea. The oldest and most complete introduction of tea in China is *The Classic of Tea* written by Lu Yu in the Tang Dynasty. The ancients recorded the origin, set and tasting methods of tea in detail, which indicates a gradually mature tea culture at that time.

The eighth chapter of *The Classic of Tea* recorded the tea producing areas of China at that time. Tianzhu Temple and Lingyin Temple were the two main tea producing areas in Qiantang area, which are the earliest records of tea in Hangzhou.

The gold belt for tea production well recognized around the world is at 30° north latitude. The belt just right passes by the West Lake—the hometown of Longjing tea. The Longjing tea cultural system is an agricultural production system centered by superior variety breeding, cultivation, brushes management, tea-leaves picking and tea culture, combined with the biodiversity it breeds, ecosystem functions it has provided and the human and natural landscape features it has presented during the production process.

Water vapors form the West Lake often smashes over the mountains southwest of the lake. Continuous hills block the coldness form the north and warmth form the south. The water vapors in the hot subtropical climate block out some of the sunlight and provide enough humidity. This is where Longjing tea bushes grow up. The emerald green tealeaves, the graceful shape, the aromatic flavor and mellow texture all benefits from the pleasant environment. "Delight flavor but waft far, aromatic fragrance but taste faint." "Longjing" was first named in the Song Dynasty, known in the Yuan Dynasty, famous in the Qin Dynasty and reputed in the Qing Dynasty. Today, the Longjing tea has worn the crown as the most famous Chinese tea.

Longjing tea is richly endowed by nature, but the sophisticated techniques also play important roles for its high reputation. There are three picking characteristics: timely, freshly and diligently. Picking when the tea leaves are tender, green and even in size is the basic requirements. After that, there are another 8 steps before a finished tea product. An experienced tea farmer can be judged from his tea-frying process: Grab, shake, jolt, rub, rub right about, push, buckle, throw, grind, and press. In the years of development, the local tea farmers have mastered a complete set of mature techniques of how to pick and fry. Techniques of planting, cultivating, picking and crafting and exquisite tea culture have inherited among the generations who devoutly love their tea gardens.

西湖龙井茶文化系统生长于浙江省杭州市。西湖畔三面环山的自然屏障所形成的独特小气候是保障龙井茶品质的重要因素，自古就是爱茶之人流连向往之处。图为清晨从空中俯瞰西湖区白乐桥，民居与茶园完美地构成了一幅田园诗画。（王艳萍 摄）

Longjing Tea Culture System is located in Hangzhou City, Zhejiang Province. West Lake is embraced on three sides by green hills, which creates an unique microclimate and ensure the high quality of Longjing tea. It has been the place where people who are fascinated with tea would linger on without any thought of leaving since long before. The picture shows in the morning overlooking the Baile Bridge in the West Lake the tea gardens and folk houses perfectly constitute a pastoral painting. Photographed by Wang Yanping

西湖龙井茶的优良品质，除得益于西湖山水的环境条件之外，最根本的是赖于西湖茶农对茶园的精耕细作、巧施用肥，及对成茶的细采精炒等一整套生产技术。龙井茶采摘有三大特点：一是早，二是嫩，三是勤。（王艳萍 摄）

Longjing tea benefits a lot from nature, but the sophisticated techniques of ploughing, fertilization, picking and frying are the keys for its high quality. There are three picking characteristics: timely, freshly and diligently. Photographed by Wang Yanping

华东地区·East China/ 浙江省·Zhejiang Province

当天采的鲜叶陆续都收来了,需要分开晾干。西湖龙井茶的炒制是为一绝,鲜叶摊放,火力锅温都需要精细掌控。龙井茶的炒制技术相当考究,手工炒制龙井茶在如今已经比较少见了。(王艳萍 摄)

The fresh leaves collected in succession on the day need to be separately dried in the air. The frying process is unrivalled, in which how to spread out the leaves and how to regulate the temperature of the pan can deceive the finial quality. But now, the sophisticated hand-fried Longjing tea is rare to be tasted. Photographed by Wang Yanping

悠久的历史和深厚的文化底蕴,让西湖龙井茶融入杭州的角角落落,吸引了无数慕名而来的游客。细细品味,或许能从一杯茶里,渐渐品出牵扯古韵遗梦的情怀。(王艳萍 摄)

The long history and profound cultural heritage make the fragrance of tea permeate into every corner of avenues and alleys, attracting countless visitors to have a taste. Just one sip of tea can call out literary feelings from the deep in the heart. Photographed by Wang Yanping

老辈人对龙井茶是敬畏的,对于这一能改变茶农生活命运的营生,自然会有很多祭拜的形式,流传至今已简化了很多。图为当地茶农在祈祷风调雨顺。(王艳萍 摄)

The elders hold the Longjing tea in awe and veneration. They make a living by that, and naturally there are many forms of worshipping ceremonies, though they have been simplified a lot so far. The picture shows that the local tea farmers are praying for good weather. Photographed by Wang Yanping

作为杭州西湖文化景观的重要组成部分，与西湖相伴相生的龙井茶文化及其茶园景观，成为地方旅游的一部分。图为当地老茶农在向景区顾客推销自家产的西湖龙井。（王艳萍 摄）

As an important part of the west lake cultural landscape in Hangzhou, Longjing tea culture and its tea garden landscape, which are associated with the west lake, have become part of local tourism. The picture shows the old local tea farmer selling their own west lake longjing to customers in the scenic spot. Photographed by Wang Yanping

低洼地区杰出的生态农业模式
浙江湖州桑基鱼塘系统

明末清初之际,中国封建社会出现了早期资本主义萌芽。其中,最具代表性的是浙江湖州的丝绸产业,该行业中的雇用劳动关系被视为中国早期资本主义萌芽的标志。

湖州丝绸产业的发展领先当时社会的发展速度,要归功于当地的农业生产模式——桑基鱼塘。

地处太湖畔的湖州,遍布常年积水的洼地。当地的农民为提高经济效益,因地制宜,把这些积水洼地改造为池塘,在池塘里养鱼,在塘基上种桑树,把桑叶拿去喂蚕,蚕的粪便蚕沙用来养鱼,鱼粪肥塘,塘泥壅桑,以此形成了一个互补共生的、良性循环的、"零污染"的农业生产系统。在这个农业生产系统中,不仅提高了土地的利用率,农民还可以得到蚕丝和鲜鱼两种产品。在农业社会,这种模式让湖州斩获了"丝绸之府、鱼米之乡"的称谓。

桑基鱼塘的生产模式在中国的太湖、洞庭湖、珠三角等河网密布、水资源丰富的地区被广泛应用,各地因农业发展速度不同,出现桑基鱼塘模式的时间也不同。成书于1149年的《农书》中写道:"在十亩地上凿陂塘二三亩,以所起之土筑堤,堤上种桑。"这是现今所能找到的最早的对桑基鱼塘的文献记载。

湖州桑基鱼塘的起源可追溯到春秋战国时期,即为太湖流域建设的塘浦圩田水利工程。为减少水患带来的困扰,人们在低洼地的四周筑堤,通过修筑"五里七里一纵浦,七里十里为一横塘"的溇港水利工程排涝防洪、引水灌溉,使"水行于圩外,田成于圩内",又在田上种桑、水里养鱼。后来,这种方式慢慢被人们发展为桑基鱼塘模式。这种模式在明清时期达到了鼎盛。进入现代社会,在多重因素的影响下,昔日规模宏大的桑基鱼塘区均日渐式微。湖州市南浔区西部的桑基鱼塘,是中国现存最集中、最大、保留最完整的桑基鱼塘系统,目前保有6万亩桑地和15万亩鱼塘。桑基鱼塘实现了对洼地的创造性利用。其最独特的价值是通过生态循环,达到了生态环境"零污染",对保护生态环境及保持经济的可持续发展,发挥了重要作用。

地理位置:北纬30°52′—30°53′、东经120°25′—120°26′,地处浙江省湖州市南浔区。

气候特点:亚热带季风气候。

认定时间:2014年被农业部列为第二批中国重要农业文化遗产,2017年被联合国粮农组织列为全球重要农业文化遗产。

Geographical location: 30°52′-30°53′N, 120°25′-120°26′E; located in Nanxun District, Huzhou City, Zhejiang Province.
Climate type: subtropical monsoon climate.
Time of identification: In 2014, it was listed as one of the second batch of the China-NIAHS by the Ministry of Agriculture. In 2017, it was listed as one of the GIAHS by the FAO.

An Outstanding Ecological Agriculture Model in Low-lying Areas
Huzhou Mulberry-dyke & Fish-pond System

The seeds of capitalism sprouted up in the feudal society in China in the late Ming Dynasty and the early Qing Dynasty. The employment relationship in the silk industry in Huzhou, Zhejiang Province is regarded as the symbol of early capitalism in China.

Thanks to the local agricultural production model—the Mulberry-dyke & Fish-pond System, the silk industry of Huzhou could lead the development of the society at that time.

Huzhou City, at the riverside of Taihu Lake, is occupied by watery low-lying lands. In order to improve economic benefits, locals deepen those low-lying lands into ponds, and raise fish in and plant mulberry trees around ponds. Mulberry leaves are used to feed silkworms, feces and silkworm excrement are used to raise fish, and the mud in ponds is used to anchor and fertilize the mulberry trees, thus forming an enclosed, cyclical, "zero" pollution agricultural production system which can not only improve the utilization rate of the land but also benefit from the silk and fish at the same time. In the agricultural society, this model help Huzhou win the title of "A Mansion Weaved by Silk, A Land Flowing with Fish and Rice".

The Mulberry-dyke & Fish-pond System is widely applied around the areas of Taihu Lake, Dongting Lake, Pearl River Delta and other areas with dense rivers and rich water resources. Because of different agricultural development speed, the system is applied in different time at different places. *The Book of Agriculture* written in 1149 recorded that "dig ponds and build its dike, then plant mulberry trees on the dike", which is the earliest literature about Mulberry-dyke & Fish-pond System.

Huzhou Mulberry-dyke & Fish-pond System can be traced back to the Spring and Autumn Period and the Warring States Period, which was the Tangpu polder water conservancy project built in the Taihu Basin. To reduce the problems caused by flooding, people built embankments around low-lying lands. The Lougang water conservation project made full preparation for flood prevention and storm drainage, then "water could run out of the polder and crops can grow in field". Besides, mulberry trees were plant on the dyke and fish was raised in the pond. Later, this farming method was slowly developed into a Mulberry-dyke & Fish-pond System which reached its peak in the Ming and Qing Dynasties.

Entering the modern society, under the influence of multiple factors, the large-scale mulberry-dyke & fish-pond area in the past has gradually declined. The mulberry-dyke & fish-pond in the western part of Nanxun District of Huzhou City is the most concentrated, largest and intact mulberry-dyke & fish-pond system in China. It currently holds 60,000 mu of mulberry land and 150,000 mu of ponds.

The mulberry-fish pond has achieved creative use of the land. What the most valuable is that it achieves "zero" pollution to the ecological environment and plays an important role in maintaining sustainable economic development by ecological cycle.

华东地区 · East China / 浙江省 · Zhejiang Province

湖州桑基鱼塘系统位于浙江省湖州市南浔区西部，这里是中国太湖南部的低洼湿地生态系统，是湖州先民顺应自然、治水兴农的智慧结晶。图为航拍湖州桑基鱼塘。（邱建申 摄）

Huzhou Mulberry-dyke & Fish-pond System is located in the west of Nanxun District, Huzhou City, Zhejiang Province. It is a low-lying wetland ecosystem in the southern part of Taihu Lake. It is also the crystallization of wisdom of Huzhou's ancestors who conformed to nature, prevent flood and flourish cultivation. Photographed by Qiu Jianshen.

"塘基种桑、桑叶喂蚕、蚕沙养鱼、鱼粪肥塘、塘泥壅桑"的生态模式是浙江湖州桑基鱼塘系统的典型特征。此系统是我国乃至世界史上人们认识、利用、改造自然的一个伟大创举,是世界传统循环生态农业的典范,是一项重要、宝贵的农业文化遗产。(邱建申 摄)

"Plant mulberry trees on the dyke, feed silkworms with leaves, provide fish nutrition with silkworm excrement, manure the pond with the feces of fish, anchor the mulberry trees with the mud from the pond", which is the typical features of Huzhou Mulberry-dyke & Fish-pond System. It is the representative of traditional circular ecological agriculture around the world, a valuable agricultural heritage and a great initiative of people to learn about, take advantage of and transform the nature world. Photographed by Qiu Jianshen.

种桑和养鱼相辅相成、桑地和池塘相连相倚，形成了江南水乡典型的桑基鱼塘生态农业景观。每年春季，当地民众便在鱼塘中育下鱼苗，到了冬季则可撒网收获。（邱建申 摄）

The mulberry field and the fish pond can supplement each other, which forms a typical ecological landscape in the south of the lower reaches of the Yangtze River. The pictures show that local farmers raise fries in spring and harvest in winter. Photographed by Qiu Jianshen.

桑基鱼塘系统是一种具有独特创造性的洼地利用方式和生态循环经济模式。其独特的生态价值实现了对生态环境的"零污染"。图为当地人采集桑叶。(邱建申 摄)

The mulberry-dyke & fish-pond system is a unique and creative way of using low-lying land and also an ecological and circular economic model. It brings "zero" pollution to the ecological environment. The picture shows locals are picking mulberry leaves. Photographed by Qiu Jianshen.

华东地区·East China/ 浙江省·Zhejiang Province

桑基鱼塘系统是人与自然和谐相处、儒家"天人合一"的"仁爱"生态伦理道德观的典范，也是体现我国道家生态哲学思想的样板。图为湖州桑基鱼塘系统养育的蚕。（邱建申 摄）

It shows a good example of the harmonious relationship between human and nature, reflects the "kindheartedness" implied in the Confucian ecological ethics that "man is an integral part of nature", and also embodies the ecological philosophy in Taoist school. The picture shows the silkworms raised by the mulberry-dyke & fishpond system. Photographed by Qiu Jianshen.

桑基鱼塘系统形成了独特的江南水乡田园景观，并且衍生出了非常丰富的文化内容。吃鱼汤饭、祭祀蚕花娘娘等，是当地的重要农事节庆活动。图为当地民众举办蚕花节，开展祭蚕神、踏青、轧蚕花、摇快船、吃蚕花饭、评蚕花姑娘等活动。（邱建申 摄）

The mulberry-dyke & fish-pond system has formed a unique rural landscape in the Yangtze River delta, and has derived a lot of related cultural customs, such as drinking rice-fish soup and worshipping the Goddess of silkworms and the like. The picture shows that the local people are holding Silkworm Flower Festival, and carrying out activities such as worshipping the god of silkworms, having an outing in spring, waving silkworm "flowers", taking a boat, eating rice growing from the system, and competing for "Silkworm Flower Girls". Photographed by Qiu Jianshen.

近年来,由于水产效益高于养蚕效益,导致重养鱼、轻养蚕,鱼塘面积增大,桑基面积缩小。为保护这一重要农业文化遗产,湖州市委、市政府正全面实施桑基鱼塘系统的保护与发展规则,促进传统桑基鱼塘生态系统的转型升级,使桑基鱼塘这一太湖边璀璨的明珠重放光彩。(邱建申 摄)

In recent years, locals pay more attention to the pond while relatively neglect mulberry fields because the benefits from raising fish are higher than the benefits from the silkworms, so the area of fish ponds has increased while that of the mulberry dykes has shrunk. To conserve the agri-cultural heritage, the Municipal Party Committee and the Municipal Government of Huzhou are implementing the conservation and development plans of the Mulberry-dyke & Fish-pond System in full scale to promote its transformation and upgrading and make the agricultural pearl at the edge of Taihu Lake shine its brightness again. Photographed by Qiu Jianshen.

世界香菇起源地
浙江庆元香菇文化系统

中国是世界上食用菌资源最丰富、最早对食用菌进行人工栽培的国家之一，目前已知的食用菌超过350种。今天餐桌上常见的人工培育食用菌香菇的发源地就在中国，距今有800多年的栽培历史。

庆元县位于浙江省西南部的丽水市，是人工培育香菇的家乡。庆元地处东南丘陵，境内群山连绵，森林覆盖率达到80%以上。同时，亚热带季风气候让庆元终年保持着合适的湿度，适宜的环境为木腐菌香菇的出现和生长提供了温床。

说到香菇，不得不提到一位关键性的人物——吴昱（因排行老三，又名吴三公，1130—1208年）。相传，庆元县农民吴三公首次对香菇进行人工栽培，并从栽培经验中总结了方法论：香菇栽培"砍花法"（也称"剁花法"）——在朽木上用斧头砍以疤痕，利用自然孢子接种栽培香菇。在这套栽培技术中，还包括选场、做樯、砍花、遮衣、倡花、开衣、惊罩、当旺、焙菇等一系列流程。庆元森林资源丰富，这套香菇栽培方法，使深山中的朽木得到了合理的利用。在这之前，香菇作为一种不太常见的野山菌，很少登上百姓的餐桌。砍花法的诞生，使香菇的命运发生历史性的转折。宋嘉熙年间（1237-1240年）之后，砍花法已流传到庆元周边一带，并且涌现出了许多以栽培香菇为业的菇民。龙泉县与庆元县相邻，在编撰于1209年的《龙泉县志》中，就有对人工栽培香菇方法的详细记录。

浙江庆元香菇文化系统经历了三个发展阶段，代表了香菇生产技术的不断革新：800多年前吴三公发明剁花法，1967年庆元利用香菇菌种栽培椴木香菇成功，1979年成立庆元县食用菌科研中心，开展袋料香菇栽培技术研究和推广。吴三公发明剁花法的伟大成就，在于它使深山老林中的"朽木"得到充分合理的利用，开创了森林菌类产品利用之先河。吴三公的发明，也使庆元成为世界香菇之源，为中国摘取了一项世界农业的桂冠。

在800多年的发展中，香菇一直是庆元的支柱性产业。现今，庆元形成了总面积1898平方千米的包括森林可持续经营、林下产业发展、香菇栽培和加工利用的农业生产系统，并以香菇为依托发展出了菇山语言"山寮白"、地方剧"二都戏"、香菇武功等丰富的香菇文化。

地理位置：北纬27°25′—27°51′、东经118°50′—119°30′，地处浙江省丽水市庆元县。

气候特点：亚热带季风气候。

认定时间：2014年被农业部列为第二批中国重要农业文化遗产。

Geographical location: 27°25′-27°51′N, 118°50′-119°30′E; located in Qingyuan County, Lishui City, Zhejiang Province.
Climate type: subtropical monsoon climate.
Time of identification: In 2014, it was listed as one of the second batch of the China-NIAHS by the Ministry of Agriculture.

The Origin of the World's Shiitake Mushroom

Qingyuan Shiitake Mushroom Culture System

China is one of the countries with the most abundant edible fungi in the world and the earliest artificial cultivation of edible fungi. Three are more than 350 edible fungi known so far. The artificially cultivated shiitake mushroom, common on the dinning-table now, is originated in China and has a cultivation history of more than 800 years.

Qingyuan County, located in Lishui City, southwest of Zhejiang Province, is the birthplace of artificially cultivated shiitake mushrooms. Located in the Southeast Hills, the rolling mountains, the vast forest covering over 80% of the land, and the stationary temperature kept by the subtropical monsoon climate create a cradle for the emergence and growth of shiitake mushrooms.

When it comes to talk about the shiitake mushroom, one person cannot be neglected—Wu Yu (1130-1208). It is said that it was Wu that cultivated shiitake mushrooms for the first time and summarized the methodology according to his cultivation experience, which was the "chopping method". Cut with an axe on the deadwood, and inoculate the shiitake mushroom with natural spores. The set of cultivation techniques also included a series of steps such as selecting an appropriate place, selecting appropriate trees, chopping, covering (the scars), uncovering... The cultivation method made reasonable utilization of the deadwoods in the vast forest of Qingyuan.

Prior to "chopping method", shiitake mushrooms, as less common wild mushrooms, were rarely served on the dinning-table. And it was the "chopping method" that brought about a historic turning point in the fate of shiitake mushrooms. After the years reigned by Emperor Jiaxi in the Song Dynasty, the chopping method was spread around the Qingyuan County. Then not a few people there began to cultivate the shiitake mushroom. Longquan County, adjacent to Qingyuan County, has a detailed record of the method of artificially cultivated shiitake mushrooms—*Records of Longquan County*, which was written in 1209.

Qingyuan Shiitake Mushroom Culture System experienced three stages of development, representing the continuous innovation of shiitake mushroom production technology: More than 800 years ago, Wu invented the chopping method; in 1967, Qingyuan County succeeded in cultivating shiitake mushrooms from basswood; in 1979, the Qingyuan Edible Fungus Research Center was established to carry out and promote researches and cultivation techniques of shiitake mushroom cultivated in plastic bag. The achievements of Wu's chopping method are that it fully and rationally utilized the "deadwood" in the deep forests, and pioneered the road for developing forest fungi products. It is also for the method that Qingyuan can be the source of the world's shiitake mushrooms, which help China win a world agricultural crown.

In the development of more than 800 years, shiitake mushrooms have always been the pillar industry of Qingyuan. Nowadays, Qingyuan has formed an agricultural production system with a total area of 1898 square kilometers, including forest sustainable management, forest industrial development, shiitake mushroom cultivation and processing and utilization technology, and developed mushroom culture such as argot spoken only by mushroom farmers, Er Du Opera, martial arts and so on.

浙江庆元香菇文化系统是庆元当地菇民秉承传统香菇种植技术，依托当地良好生态环境和丰富森林资源，从事香菇生产而形成的农业文化系统。图为庆元当今最大的香菇基地的全景图。（郑承春 摄）

Qingyuan Shiitake Mushroom Culture System is an agri-cultural system adhering to traditional shiitake mushroom cultivation technology, relying on local favorable ecological environment and rich forest resources. The picture shows the panorama of the largest mushroom base in Qingyuan today. Photographed by Zheng Chengchun

"香菇寮"是庆元早年菇民的必备空间。据称，那个时候从农历九月末开始，菇民们便要离妻别子、背井离乡，奔赴遥远的菇寮，在那里开始漫长的种菇生活：头两年冬春季为砍树种菇的季节，第三年冬春季开始收成，一般能收三至四年。菇民在选定的山场上，取坐北朝南、通途有道、取水方便、地形隐蔽之处，以竹、木、茅草为材料，搭建简易寮棚，俗称"香菇寮"。图为20世纪80年代的"香菇寮"和留存至今的"香菇寮"。（上图 姚家飞 摄；下图 王斌 摄）

Thatching sheds were necessary for shiitake mushroom grower in the early years. It was alleged that, from the end of September of the lunar calendar, the mushroom people had to leave their hometown and family to go to distant mushroom fields, where they began a long life of mushroom farming: winter and spring in the first two years were the time for cutting trees and growing mushrooms, then harvest in the same seasons in the third year. The harvest period could last for three or four years. They selected secluded area where it faced south, had convenient transportation and sufficient water; and then built makeshift thatching sheds, where were also nicknamed "shiitake mushroom room". The pictures show the "shiitake mushroom room" in the 1980s and some that have survived to this day. The picture above photographed by Yao Jiafei, and below by Wang Bin

剁花法使深山老林中的"朽木"得到充分合理的利用，开创了森林菌类产品利用之先河。菇农用原木培植，在树木上砍出花来，然后依靠天然香菇孢子传播接种，虽然生长周期长，但是肉厚味鲜。图为当地菇农在用剁花法种植香菇、生长中的椴木香菇及采摘成菇。（王斌、郑承春 摄）

The chopping method fully utilized the "deadwood" in the deep mountain forests, which pioneers the use of forest fungi products. The shiitake mushrooms grow on the round timber: first chop, then inoculate with natural shiitake mushroom spores. Though the growth cycle is relative longer, the pileus is thicker and more delicious. The pictures show the local mushroom farmers are chopping woods and picking shiitake mushrooms; growing basswood mushrooms. Photographed by Wang Bin and Zheng Chengchun

20世纪90年代以后，袋料培植香菇的技术逐渐在庆元普及，香菇产量获得大幅提高，尤其是高棚层架栽培袋料香菇，既提高了香菇品质，又节省了制菇占用耕地。图为袋料高棚制菇、采菇。（郑承春 摄）

After the 1990s, the technology of bagging cultivation of shiitake mushrooms gradually became popular in Qingyuan, and the field of shiitake mushrooms was greatly improved. Especially the techniques of high-shelf cultivated shiitake mushrooms in plastic bags, it not only improved the quality of shiitake mushrooms, but also saved the cultivated land occupied by mushrooms. The picture shows the high-shelf cultivated shiitake mushrooms in plastic bags and farmers picking mushrooms. Photographed by Zheng Chengchun

庆元香菇采摘后，经过自然晾晒或人工烘干等工序，可以获得很好的保存。（郑承春、张用权 摄）

After harvest, Qingyuan shiitake mushrooms can be well preserved after natural drying or manual drying. Photographed by Zheng Chengchun and Zhang Yongquan

华东地区·East China/浙江省·Zhejiang Province

吴三公,庆元县百山祖镇龙岩村人,因排行第三,俗名"吴三",又因发明人工栽培香菇技术,被菇民尊为菇神,称"吴三公"。他去世后,百姓为纪念他,建庙祭祀,逐渐演化为神。(王斌 摄)

Wu Sangong, a native of Longyan Village, Baishanzu Township, Qingyuan County, was ranked third, and so his nickname was Third Wu. Also because he invented the artificial cultivation method of shiitake mushrooms, Wu was honored as the god of mushroom or muster Wu. After he passed away, people built temples, worshipped and regarded him as a celestial being. Photographed by Wang Bin

西洋殿,又名"松源殿""吴判府殿",坐落于庆元县五大堡乡西洋村松源溪畔,系古代菇民为纪念香菇鼻祖吴三公而建的纪念性建筑,始建于宋咸淳元年(1265年)。图中远景建筑即西洋殿——香菇鼻祖吴三公庙。(陈士平 摄)

Xiyang Temple, also known as "Songyuan Temple" and "Palace of Master Wu", built in the first year reigned by Emperor Xianchun in the Song dynasty (1265) at the riverside of Songyuan River, Xiyang Village, Five Township, Qingyuan County, is a monumental building built by the ancient mushroom farmers to commemorate the originator of the mushroom–Wu Yu. The distant view in the picture is the Xiyang Temple. Photographed by Chen Shiping

每年农历三月十七日（吴三公生日）和七月十六日至十九日是龙泉、庆元、景宁三县菇民朝拜祀奉"菇神"的进香期，在此期间会举行菇神庙会。图为西洋殿内香菇鼻祖吴三公朝圣大典祭祀的情景。（郑承春 摄）

A temple fair is held each year on the 17th day of March (the birthday of Wu) and one the 16th and 19th days of July in the lunar calendar, people in Longquan, Qingyuan, Jingning counties to worship and pay respects to the "god of shiitake mushroom". The picture shows the grand ritual ceremony. Photographed by Zheng Chengchun

"二都戏"是起源于庆元县的一种具有浓郁地方特色的稀有的传统戏曲剧种。由于庆元左溪、荷地、合湖一带古称"二都",因此得名。二都戏表演者大多为菇民自身,而且其传承、交流和发展均与菇民的生产生活习俗密切相关,故又被称为"菇民戏"。图为当地人在西洋殿观看二都戏。(郑承春 摄)

The Er Du Opera is deeply characterized by the Qingyuan culture. It is named after the place"Er Du", the combination of Zuoxi, Hedi, Hehu townships. The opera is also known as the "Drama of Shiitake Mushroom Farmers" because most of the performers are local farmers, and the inheritance and development of the opera are closely related to the customs of their life and production. The picture shows the locals are enjoying the Er Du Opera in Xiyang Temple. Photographed by Zheng Chengchun

庆元菇民世代在深山老林中劳作，形成了包括菇山语言"山寮白"、地方剧"二都戏"、香菇武功等在内的绚丽多姿的香菇文化。图为香菇武功练功场景。（郑承春 摄）

The toilers hardworking in deep forests create and develop rich culture centered by the cultivation of shiitake mushroom such argot spoken only by mushroom farmers, Er Du Opera, martial arts and so on. The picture shows the martial art. Photographed by Zheng Chengchun

菇木林是指用于香菇剁花法栽培的林木。为保持菇木林的可持续经营，其面积从百多亩到上千亩不等，通常采用"自下而上的分区作业，伐大留小的择伐方式"，实现了林木的可持续经营。（王斌 摄）

Mushroom trees in Qingyuan County refers to the trees used for mushroom cultivation by chopping method. In order to maintain the sustainable management of mushroom forests, the area of mushroom forests ranges from more than 100 mu to more than 1000 mu. Usually, "the felling model of cutting from the bottom zone to the top zone and cutting the large trees while retaining the small ones" is adopted to realize the sustainable management of mushroom forests. Photographed by Wang Bin

湿地山地立体农业生产体系
福建福州茉莉花与茶文化系统

福州是传统的茶叶种植区。闽江穿城而过,给福州留下了独一无二的河口盆地。盆地四周环绕着海拔600—1000米的山地,日照短,多散射,云雾缭绕,湿度大,土质是酸性的红壤,满足了茶树生长所需的一切条件。当地的方山露芽、鼓山半岩茶、罗源七境绿茶在历史上均是有名的贡茶。

汉代,海上丝绸之路形成,茉莉花远渡重洋,"随经入汉京",来到了港口城市东冶(今福州)。福州地处亚热带季风气候区,气候湿润、热量充足,沿闽江的平原地区土壤肥沃,扦插茉莉易成活。茉莉花这一原产于中亚细亚的植物,在福州扎下了根。茉莉花姿态素雅、香气清芬,与中国人所追求的君子之风气质十分吻合,很快便获得了人们的喜爱。

北宋年间,福州已经是茉莉花满城飘香,形成了"山丘栽茶树,沿河种茉莉"的种植格局。《瓯冶遗事》有记载:"果有荔枝,花有茉莉,天下未有。"福州至今仍保留着北宋年间福州太守蔡襄的题刻"天香台",这里的"天香"就是指茉莉花。

彼时,中医认为香花与茶叶结合,对人体的保健有着独特作用。受这一观点的影响,南宋时用香花加工花茶的工艺开始兴起,茉莉花茶的窨制工艺开始被福州人尝试和改进,产生了独特的茉莉花茶,并随时代发展,日臻成熟。

福州出产的茉莉花茶,凭借清新的香气在花茶中独树一帜,获得了市场的认可。随着时间的推移,炮制茉莉花茶的技艺日渐成熟,茉莉花茶在福州逐渐形成产业,并形成了适应当地生态条件的茉莉花基地(湿地)—茶园(山地)的循环有机生态农业系统,既保持了生态系统的生物多样性,又提高了单位面积的生产效益。

茉莉花茶的炮制严格采用春茶和伏花,窨制工艺包括茶坯处理、鲜花养护、茶花拌和、堆窨、通花、收堆、起花、烘焙、冷却、转窨或提花、匀堆装箱等数十道工序。茉莉花茶的窨制工艺,至今没有外传,使得茉莉花茶成为我国独有的茶叶品种。

地理位置:北纬25°15′—26°39′、东经118°08′—120°31′,地处福建省福州市。
气候特点:亚热带季风气候。
认定时间:2013年被农业部列为首批中国重要农业文化遗产,2014年被联合国粮农组织列为全球重要农业文化遗产。

Geographical location: 25°15′-26°39′N, 118°08′-120°31′E; located in Fuzhou City, Fujian Province.
Climate type: subtropical monsoon climate.
Time of identification: In 2013, it was listed as one of the first batch of the China-NIAHS by the Ministry of Agriculture. In 2014, it was listed as one of the GIAHS by the FAO.

The Stereoscopic Agricultural Production System in Wetlands and Mountainous Areas
Fuzhou Jasmine and Tea Culture System

Fuzhou is a traditional tea growing region. The Minjiang River passes through the city, leaving Fuzhou with a unique Estuarine Basin. The mist-shrouded basin, surrounded by mountains at an altitude of 600-1000 meters, combined with fewer hours of sunshine, mainly scattering, and high humidity and acid red soil, provide a perfect environment for the growth of tea plants. Dew bud tea from Fangshan Mountain, Banyan tea from Gushan Mountain and green tea from Qijing, Luouyan County are all well-known tributes in the long history of China.

In the Han Dynasty, the Jasmine travelled on the maritime Silk Road across oceans and arrived at the port city Dongye (now Fuzhou City). Fuzhou is located in a subtropical monsoon climate zone. With the humid climate, abundant heart and fecund soil of plains along the Minjiang River, the jasmine planted by cuttage method was easy to survive. Since then, the jasmine, a plant native to Central Asia, rooted in Fuzhou. The jasmine is elegant and graceful in posture and light in fragrance, which is very accordance with the gentleman's temperament pursued by the Chinese, and it quickly became popular in China.

During the Northern Song Dynasty, Fuzhou was already a city of jasmine flowers. The crop planting pattern that "Plant tea trees on hills and jasmine along the riversides" was formed. "Fuzhou's people have already tasted the juicy litchis and appreciated the grace of jasmines which were still unknown in other places in the world", recorded in the book *Enterprises of Ou Ye*. The pavilion "Fragrance from Paradise", which was name by Cai Xiang, the prefecture chief of Fuzhou in the Song Dynasty, stands still in Fuzhou today. The "fragrance" here refers to the fragrance of jasmines.

At that time, doctors of Chinese medicine believed that the combination of fragrant flowers and tea had a unique effect on human health. Enlightened by the view, scented tea began to rise in the Southern Song Dynasty. Fuzhou people improve the scenting craft to make characteristic fragrant jasmine tea, which become better and approaching perfection day by day.

The jasmine tea produced in Fuzhou has earned a place in the scented tea market with its special fresh aroma. As time goes on, jasmine tea was industrialized in Fuzhou by applying the mature techniques. A circular organic ecological agricultural jasmine-base(wetland) & tea-garden (mountain)system that adapts to the local ecological conditions is formed, which not only maintains the biodiversity of the ecosystem, but also improves the production efficiency per unit area.

Siring and mixing, piling, baking, cooling...there are ten steps for making the jasmine tea. The scenting craft is only available for China, making jasmine tea of Fuzhou a unique tea variety in China.

福建福州茉莉花种植与茶文化系统是以在江边沙洲种植茉莉花、在高山上种植茶叶为特色的湿地茉莉花山地茶园相结合的生态农业系统。遗产地位于福州市的6个县市区，现有茉莉花种植面积约1.5万亩，茶园面积约13.5万亩。图为今天的福州恩顶茶园。（林耘 摄）

Fuzhou Jasmine and Tea Culture System is a circular ecological agricultural system applied in featured jasmine planted in sandbank along riversides and tea trees planted on hills. The heritage fields are distributed in Fuzhou's 6 counties, cities and districts. Now, there are about 135,000 mu of jasmine planting area and 135,000mu of tea gardens. The picture shows the En Ding tea garden in Fuzhou. Photographed by Lin Yun

茉莉花源于中亚细亚，茶源于中国，它们的结合是2000年东西方文化交流的见证。福州茉莉花种植与茶文化系统在长达2000年的协同进化过程中逐渐完善，是古人利用环境、适应环境发展农业的典范，是农业的活化石。图为福州白云村茉莉园和雪峰高山茶场。（林耘 摄）

The jasmine, native to Central Asia, and tea, native to China, witness the two thousand years of cultural exchanged between East and West. Fuzhou Jasmine and Tea Culture System has gradually improved in the process of two thousands of years of co-evolution. It is an example set by the ancients to utilize and adapt the environment to develop agriculture; it is the living fossil for researching agriculture. The pictures show the jasmine garden in White Cloud Village and the tea plantation in Snow Mountain Village in Fuzhou. Photographed by Lin Yun

福州茉莉花采摘期在 5 月至 10 月，三伏天的茉莉花开得最大最香，是窨制茉莉花茶最好的花料。采茉莉花需要在晴朗的天气下进行，采摘时间一般选择一天中最热的时候，从中午 12 点到下午 4 点，这个时候的花形态最丰满。茶叶的采摘也很讲究，不仅讲究手法，天气一般也需要晴天。（上图 张永艳 摄；下图 林耘 摄）

Jasmine flowers picked from May to October each year, especially in dog days, are the best materials for scenting the tea. The hottest time period on sunshiny days, usually from 12 am to 4 pm when flowers bloom the most, is the best time for picking jasmine. The sunshiny weather also plays a deceive role in the high quality of tea. The picture above photographed by Zhang Yongyan, and below by Lin Yun

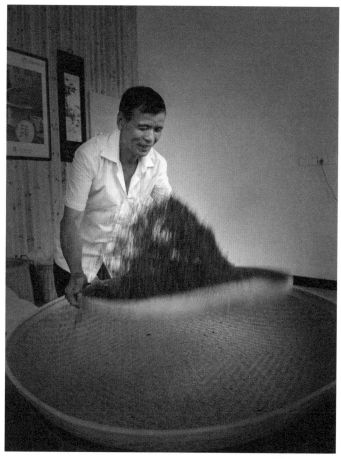

茉莉花茶是中国独一无二的茶叶品种。茉莉花茶采用春茶伏花原料,窨制工序有数十道。由于历史上福州人对工艺严格保密,窨制工艺在数百年间均未传到其他国家,目前世界上只有中国能窨制茉莉花茶。图为茉莉花茶的采摘和加工窨制过程。(林耘 摄)

Jasmine tea is a unique tea variety in China. The tea picked in spring, the flowers picked in summer and the making craft including dozens of meticulous steps, all make what the jasmine is. Because Fuzhou people have kept their craftsmanship as a secret for hundreds of years, the scenting craft has not been transmitted to other countries. At present, only China holds the scenting method in the world. The pictures show the processing scenting of jasmine tea. Photographed by Lin Yun

由于城市建设和其他产业发展,福州茉莉花茶传统生产模式变得濒危,花茶窨制工艺面临着严峻考验。近年来,由于福州市委、市政府的重视和保护,福州茉莉花茶这一重要的传统农业产品,正以全新的姿态,欢迎世界各地的朋友来共同品鉴。图为农业部国际合作司 GAP 茉莉花茶出口生产基地。(林耘 摄)

Due to urban construction and other industrial developments, the traditional production mode of jasmine tea in Fuzhou has become endangered, and the scenting craft is facing a severe test. In recent years, Fuzhou Municipal Party Committee and the Municipal Government are paying more attention to the traditional cultural product. The jasmine tea, with a new look, is welcoming friends from all over the world to come and have a taste. Photographed by Lin Yun

竹林、村庄、梯田、水系综合利用模式

福建尤溪联合梯田

我国秦岭—淮河以南、青藏高原以东的广大地区，遍布着高低起伏、连绵不断的山地和丘陵。在热带和亚热带季风气候的影响下，这些地区往往降水丰富、热量充足。在农业生产中，喜热喜湿的水稻当仁不让地成为主要农作物。

但是，广布丘陵、山地的地形，并不能满足水稻种植所需要的大面积平整土地的条件。在山地丘陵上修筑梯田，成为生活在南方山区的农民解决问题的关键。此外，除了可以解决土地需求外，修筑梯田对于一些山体坡度大、降水强度大的地区来说，还有着保持水土的作用。

中国的梯田主要分布在南方一些多山少地的丘陵地带，其中以广西、福建、江西、贵州、湖南和云南居多。依照地形、沿山坡坡面修成的连片梯田，几乎成为南方传统稻作区的标志。

福建省三明市尤溪县地处戴云山脉北段西部，山地和丘陵地貌占到了县域总面积的93%，很少有大面积的平整土地。尤溪县的联合乡有一片保存完整、面积达一万多亩的梯田。这片梯田跨8个行政村，绵延几十里，是中国历史上开凿最早的大型古梯田群之一。历经几百年的发展，梯田铺满了联合乡大大小小的山坡高地。梯田中面积最大的田块有数亩大，最小的仅如簸箕般大小，位置最高的田块海拔近900米，最低的有海拔260多米。在一座山坡中，梯田最高级数可达上千级。

自唐开元二十九年（741年）尤溪建县起，当地居民就已经开始在山坡上开垦梯田。至今，梯田已占到尤溪耕地总面积的95%。

尤溪境内水资源丰富，除自然降水外，还有大大小小34条河流，给当地的农业生产提供了充足的水量。联合乡的农民通过山顶的竹林将自然降水留住，再通过溪流使水流入梯田和村庄，为农田灌溉和农民生活所用，形成了联合乡独特的"竹林—村庄—梯田—水流"山地农业体系。

地理位置：北纬25°50′—26°26′、东经117°48′—118°40′，地处福建省三明市尤溪县。

气候特点：亚热带湿润性季风气候。

认定时间：2013年被农业部列为首批中国重要农业文化遗产，作为"中国南方山地稻作梯田系统"之一，于2018年被联合国粮农组织列为全球重要农业文化遗产。

Geographical location: 25°50′-26°26′N, 117°48′-118°40′E; located in Youxi County, Sanming City, Fujian Province.

Climate type: subtropical humid monsoon climate.

Time of identification: In 2013, it was listed as one of the first batch of the China-NIAHS by the Ministry of Agriculture. In 2018, as one of the Rice Terraces in Southern Mountainous and Hilly Areas in China, it was listed as one of the GIAHS by the FAO.

A Comprehensive Utilization Model of Bamboo Forests, Villages, Terraces and Rivers System
Youxi Lianhe Terraces

Undulating and continuous mountains and hills spread across the south of Qinling Mountains-Huaihe River line and the vast area to the east of the Qinghai-Tibet Plateau. The tropical and subtropical monsoon climate, bring abundant rainfalls and sufficient heat. In agricultural production, rice, a crop that enjoys heat and moist, naturally take up its responsibility as a main crop.

The mountainous landform, however, cannot meet the requirements for large-scale planting area. How to build terraces on mountains and hills becomes the first problem to be solved. Building terraces cannot only satisfy the land demand, but also can preserve water and soil for some area with steep slopes and high precipitation intensity.

China's terraced fields are mainly distributed in hilly areas, especially in Guangxi ,Fujian,Jiangxi,Guizhou,Hunan and Yunnan. These continuous terraces built based on different landform and gradient have become a symbol of the traditional rice-growing area in the south.

Youxi County, Sanming City, Fujian Province is located in the western part of the northern section of the Daiyun Mountain Range. The mountainous and hilly landforms account for 93% of the total area of the county, so there is very little smooth land suitable for planting. There is a terrace with an area of over 10,000 mu lying in Lianhe Township, Youxi County. It spans 8 administrative villages and stretches for dozens of miles, one of the earliest large-scale ancient terraced fields in the history of China. After hundreds of years of development, mountains and highlands of Lianhe Township are overspread with terraces. The largest terraces can cover an area of thousands of square meters, while the smallest is only the size of a dustpan; the highest is nearly 900 meters while the lowest is about 260 meters. Some can have more than 1,000 stairs.

Since Youxi was built in the Kaiyuan flourishing age in the Tang Dynasty (741), local residents had begun to bring mountain and slopes under cultivation. Now, terraces account for 95% of the total area of cultivated land is in Youxi.

Youxi has abundant water resources. Natural rainfalls and 34 rivers, large or small, provide sufficient water for local agricultural production. Rainfall is preserved by the bamboo forest on the top of the mountain, and then flow into rivers to be used for irrigation or drinking, forming a unique "bamboo forests-villages-terraces-rivers" agricultural system applied to mountainous area.

福建尤溪联合梯田是以南方中低山丘陵地区梯田景观为特色的稻作梯田生态系统。自唐宋以来，联合村民使用木犁、锄头等工具开垦梯田、种植水稻，在险峻的金鸡山中创造了神奇壮丽的梯田，成为村民几百年来的主要生产生存方式。（包世生 摄）

Youxi Lianhe Terraces forms a rice terraced ecosystem characterized by terraced landscapes in the middle and low hills of the south China. Since the Tang and Song Dynasties, Lianhe villagers used wooden plows, hoes and other tools to build terraces and plant rice, creating magical and magnificent terraces in the steep Jinji Mountain, which has been the main survival mode of Lianhe people for hundreds of years. Photographed by Bao Shisheng

联合梯田依山开垦，形成形态丰富的立体生态景观，通过山顶竹林截留、储存天然降水，再以溪流流入村庄和梯田，形成特有的"竹林—村庄—梯田—水流"山地农业体系。（包世生 摄）

Lianhe terraces built based on mountainous terrain create a three-dimensional ecological landscape. Besides, Rainfall is preserved by the bamboo forests on the top of the mountain, and then flow into rivers to be used for irrigation or drinking, forming a unique "bamboo forests-villages-terraces-rivers" agricultural system applied to mountainous area. Photographed by Bao Shisheng

春天，农民给田里灌水浸烂田泥；到插秧时，农民种上田埂豆、放些鱼苗，鲤鱼能减少田中杂草的生长和虫害的发生，田埂豆发达的根系能保护田埂；收获时，再放干田里的水，收鱼、收水稻、收黄豆。收获后，鸭子、山羊等被赶入田中，觅食遗撒的谷粒和新长出的杂草。动物粪便、作物秸秆和豆类的固氮功能，则使土壤肥力不断提升。丰富的生物多样性和复合生产模式可以降低病虫害发生率，维持梯田生态系统的平衡。（包世生 摄）

In the spring, irrigate terraces for immersing soil; when transplanting rice seedlings, plant soybeans on the dykes and put juvenile fish into terraces; when it is time to harvest, drain the water and then harvest fish, rice and soybeans. After the harvest, let ducks, goats, etc. into terraces to eat left grains and weeds newly growing. Soybeans can anchor the soil and carps can eat weeds and pests. Besides, the soil fertility can be continuously improved by animal manure, crop straws, and the nitrogen fixation of soybeans. Various species and compound production models reduce the incidence of pests and diseases and maintain the balance of the terrace ecosystem. Photographed by Bao Shisheng

梯田垂直落差 600 多米，绵延数十里，田在山中，群山环抱，土墙灰瓦的村落散落其间，一派与世无争的安然祥和。（包世生 摄）

Small villages sleep soundly in the arms of mighty mountains, companied by the terraces extending for several miles with more than 600 meters of vertical drop. All are in quiet and peace. Photographed by Bao Shisheng

联合梯田规模宏大、气势磅礴，走入联合梯田，便仿佛走入了一个神奇的世界。梯田层层叠叠，犹如流云飞波，线条优美，疏密相间，造型各异，构成一幅幅山水田园诗画。（包世生 摄）

The grand and magnificent Lianhe terraces make you feel walking into a magical world. The stacked terraces are like flying clouds in beautiful lines, different shapes, drawing a pastoral landscape painting. Photographed by Bao Shisheng

这里的气候也十分特异，常常云海绵绵、雾涛滚滚，美丽的梯田、青山在云雾中忽隐忽现，闪烁着亮光，像一面面神女的镜子，熠熠生辉。这一切组合在一起，形成了神奇壮丽的景观。（包世生 摄）

Lianhe enjoys specific climate. Terraces often flicker in the rolling clouds, shining like mirrors of goddess. All of this combined to form a magical and magnificent landscape. Photographed by Bao Shisheng

尤溪联合梯田保留了多样的节庆文化与农耕习俗，如鞭牛迎春、耕牛节、丰收节、歇冬节等蕴含农耕时节传统文化的庆典仪式。其中立春鞭牛据传源于唐宋，当天一早，地方官须身着官服，手执彩杖，面向土牛念"祈福祝文"，后绕牛一周，用彩杖击打牛身三次，而后百姓围上去，用棍把土牛打成土屑，儿童则唱童谣。仪式结束，百姓竞相把细土带回家，洒在自家地里或院子里的树木花草中，表示接到春、接到福。（包世生 摄）

Youxi Lianhe Terraces retain various festivals and customs, such as celebrating the spring with the whip of the cattle, the ploughing cattle festival, the harvest festival and the winter festival of rest. Take celebrating the spring with the whip of the cattle which can be traced back to the Tang and Song dynasties for example, in the early morning of the day of spring, local officials were required to wear official clothes, hold colorful sticks, and read "blessing and blessing" to the local cattle. After walking around the cattle, they hit the cattle with colored sticks three times. Then the people surrounded up, with sticks to beat the cattle into dust, children are singing nursery rhymes. After the ceremony, the people race to bring home the fine soil, sprinkle in their own fields or yard in the trees and flowers, said to receive spring, received blessing. Photographed by Bao Shisheng

相传联合乡的先民在开垦梯田的过程中，山岭险恶，猛虎为患，伏虎禅师帮助联合先民降服猛虎，克服困难，为联合乡的先民带来风调雨顺。为了答谢这位禅师，每年的农历二月廿七日，联合乡都会举办伏虎岩庙会，邻近地区群众近万人参与。每年农历十二月廿三前后，联合百姓都会举办"歇冬节"，感谢五谷仙送来风调雨顺，感谢耕牛一年的辛苦劳作。图为俯瞰伏虎岩庙会和联合乡云山村歇冬节祭拜五谷仙和牛神的场景。（包世生 摄）

It was said that in the process of reclaiming terraces the ancestors of lianhe had be helped by buddhist monk to fight tiger attacking, overcome difficulties, and bring good weather. In order to thank this monk, every year on the 27th day of the 2nd month of the lunar calendar, Lianhe Township will hold Fuhuyan Temple Fair, which was attended by nearly ten thousands of people from the neighboring areas. Every year around the 23rd day of the 12th lunar month, the Lianhe people will hold the"Winter Festival of Rest"to thank the grain fairy for bringing good weather, and thank the cattle for a year's hard work. The pictures show the scenes Fuhuyan Temple Fair and the"Winter Festival of Rest"in Yunshan Village of Lianhe Township. Photographed by Bao Shisheng

华东地区·East China/福建省·Fujian Province

乌龙茶发源地
福建安溪铁观音茶文化系统

中国饮茶的历史悠久，茶叶品类丰富。根据制作方法和发酵程度，可以将茶叶分为六大类：绿茶、白茶、黄茶、乌龙茶、黑茶和红茶。其中，"最年轻"的茶类当属乌龙茶。

乌龙茶是介于不发酵的绿茶和全发酵的红茶之间的半发酵茶，工艺复杂、制作费时、炮制考究，出现在中国茶叶产业已经发展得相当成熟的明末清初之际。今天，我国的广东凤凰、闽南、闽北、台湾台北等地区都是著名的乌龙茶产区。查究这些地区乌龙茶的起源，都离不开一个地方——福建安溪。

安溪位于中国茶叶大省福建的东南部，有"闽南茶都"的美誉。根据《安溪县志》记载，这里唐朝末年开始生产茶叶，当时在小范围内已小有名气。宋代时，安溪的茶叶产业发展达到了当时社会的较高水平。此时，安溪茶叶开始通过泉州港沿海上丝绸之路出口到东南亚等地区。

到了明清时期，中国茶叶的发展达到鼎盛，制茶技艺有了新的突破，安溪茶叶的发展也随之进步。1725年前后，安溪的茶人在以往的制茶工艺中，加入了一道新的工序——包揉。包揉法的使用，诞生了一种新的茶类——半发酵的乌龙茶。福建安溪铁观音茶文化系统，就是以种植茶叶和制作铁观音茶为特色的农业生态及茶文化系统。安溪是铁观音的发源地。铁观音既是茶叶名称，又是茶树品种名称。这一品种于清雍正年间（1723—1735年）在福建省安溪县西坪镇发现并开始推广铁观音。

彼时，安溪的茶人用乌龙茶的制法，来炮制铁观音茶树的叶子，由此产生了"铁观音茶"。采用包揉法制作的半发酵茶叶铁观音味道香浓、沁人心脾，甫一上市，便取得了市场的认可，很快传遍了台湾等乌龙茶区。

铁观音茶的采制，分为"看青做青"和"看天做青"两种技法，包括"三大阶段十道工序"，因其复杂性和对人工技法的高要求，被誉为"最高超、最精湛、最独特的制茶技艺"，2008年入选国家级非物质文化遗产保护名录。历经百年发展，安溪县形成了以传统铁观音品种选育、种植栽培、植保管理、采制工艺和茶文化为核心的农业生产系统，安溪人的生活早已与铁观音茶融为一体。

地理位置：北纬24°50′—25°26′、东经117°36′—118°17′，地处福建省泉州市安溪县。

气候特点：亚热带湿润性季风气候。

认定时间：2014年被农业部列为第二批中国重要农业文化遗产。

Geographical location: 24°50′-25°26′N, 117°36′-118°17′E; located in Anxi County, Quanzou City, Fujian Province.

Climate type: subtropical humid monsoon climate.

Time of identification: In 2014, it was listed as one of the second batch of the China-NIAHS by the Ministry of Agriculture.

The Birthplace of Oolong Tea
Anxi Tie Guanyin Tea Culture System

China enjoys rich tea varieties and a long tea-drinking history. According to the production method and the degree of fermentation, tea can be divided into six categories: green tea, white tea, yellow tea, oolong tea, dark tea and black tea. Among them, Oolong tea is the youngest one.

Oolong tea is a semi-fermented tea, intermediate between non-fermented green tea and fully fermented black tea. Its complicated, time-consuming, and high demanding process appeared in the late Ming and early Qing dynasties when the tea industry was highly developed and held mature producing techniques. Today, Phoenix Township in Guangdong Province, Southern and Northern Fujian Province, Taipei of Taiwan Province are all famous oolong tea producing areas. For tracing the origin of oolong tea of these areas, a place cannot be omitted—Anxi County in Fujian Province.

Anxi is located in the southeastern part of Fujian Province, a major tea producing province in China. Anxi enjoys the reputation of "The Tea Capital of Southern Fujian". According to the *Records of Anxi County*, Anxi began to produce tea in the late Tang Dynasty, and the tea produced here attained some renown. In the Song Dynasty, the development of the tea industry in Anxi reached a relatively high level. At this time, Anxi Tea began to be exported through Quanzhou Port along Maritime Silk Road to the Southeast Asia and other regions.

In the Ming and Qing Dynasties, the development of Chinese tea reached its peak, and the tea-making skills had a new breakthrough. The development of Anxi tea also improved. Around 1725, a new step was added into the tea making process—bags rubbing. It was the bags rubbing that gave birth to a new type of tea—fermented oolong tea.

Anxi Tie Guanyin Tea Culture System is an agroecological and tea culture system featuring Tie Guanyin Tea planting and making. Anxi is the birthplace of Tie Guanyin Tea. Tie Guanyin Tea is the name both of the tea and the tea bush. It was discovered in Xiping Township, Anxi County, Fujian Province during the years reigned by Emperor Yongzheng in the Qing Dynasty.

At that time, Anxi people used the making method of oolong tea to process the leaves of Tie Guanyin, and then produced the Tie Guanyin tea. The semi-fermented Tie Guanyin, which was made by the making methods including the bags rubbing step, was aromatic and refreshing. It was quickly accepted and recognized by the market and quickly spread to Taiwan and other oolong producing areas.

The tea-picking and processing step is divided into two techniques: according to the color of tea leaves and according to the "color" of the weather. For the complex and high demanding processing, including three major stages and ten steps, the tea making process of Tie Guanyin is known as the "the most exquisite, the most consummate, and the most unique" techniques. In 2008, it was listed into the National Intangible Cultural Heritage Protection List. After a hundred years of development, Anxi County has formed an agricultural production system with breeding, cultivation, tea protection and management, picking and processing techniques and tea culture as its core. The life of Anxi people has already merged with Tie Guanyin tea.

福建安溪铁观音茶文化系统位于福建省东南部晋江西溪上游的安溪县西坪镇，居山而近海。核心区位于安溪县西坪镇，包括松岩、尧山、尧阳、上尧、南阳5个村。该区春末夏初雨热同步，秋冬两季光湿互补，十分适宜茶树生长。图为当地茶山风光。（李子宏 摄）

Anxi Tie Guanyin Tea Culture System is located in Xiping Town, Anxi County, southeast Fujian, on the upper reaches of the western branch of Chin River, and close to mountains and the sea. There are five major tea planting villages: Songyan Village, Yaoshan Village, Yaoyang Village, Shangyao Village and Nanyang Village, all were located in Xingping Township, Anxi County. In the late spring and early summer, rain and heat will come together; in autumn and winter, the sunlight and humidity come on the stage, which is very suitable for the growth of tea trees. The picture shows the local mountain scenery. Photographed by Li Zihong

福建安溪西坪镇是闻名遐迩的铁观音发源地,极为宝贵的铁观音母树就生长在西坪镇的观音山上,仅有两棵,树龄超过 200 年。(闵庆文 摄)

Xiping is a well-known birthplace of Tie Guanyin. The only two precious seed trees, over 200 years old, grow on the Guanyin Mountain in Xiping. Photographed by Min Qingwen

安溪铁观音的制作综合了红茶发酵和绿茶不发酵的特点，属于半发酵的品种，采回的鲜叶力求完整。鲜叶按标准来收进厂，经过凉育后进行晒青。晒青时间以午后4时阳光柔和时为宜，叶子摊薄，以失去原有光泽，叶色转暗，手摸叶子柔软，顶叶下垂，失重6%—9%左右为适度。图为采青、晒青。（李子宏 摄）

The production of Anxi Tie Guanyin combines the characteristics of black tea with fermentation and green tea without fermentation, belonging to the semi-fermentation. The fresh leaves harvested are sought to be complete. Fresh leaves are collected into the factory according to the standard and dried when they get cooler. The best time of drying is after 4 o'clock in the afternoon when the sunlight is gentle. Straight and spread out the leaves until lose the original luster and turn into dim green. When leaves get soft, the top leaves droop and the weight loss about 6%-9%, then the dying is down. Photographed by Li Zihong

据《安溪县志》记载：安溪产茶始于唐末，兴于明清，盛于当代，至今已有1000多年的历史。千百年来，安溪人在长期实践中，发明出"三大阶段十道工序"的传统制作技艺，被茶业界誉为"最高超、最精湛、最独特的制茶技艺"。图为凉青、摇青、杀青、包揉等铁观音的采制过程场景。（李子宏 摄）

According to *Records of Anxi County*, Anxi tea began in the late Tang Dynasty and flourished in the Ming and Qing Dynasties. It has a history of more than 1,000 years. For thousands of years, Anxi people have invented the traditional production techniques including "three major stages and ten steps" in the long-term practice, which is praised by the market as the "most exquisite, the most consummate, and the most unique" techniques. The pictures show the steps of cooling, rolling, crank and bags rubbing and so on. Photographed by Li Zihong

安溪县茶农还发明出"茶叶短穗扦插技术",不仅让安溪成为全国茶树无性繁殖发源地,也可以通过这种繁殖方式扩大生产,让世界各地的人都能品尝到品种纯正、品质优良的铁观音。(李子宏 摄)

Anxi people also invent the "tea-branch cuttage techniques", which not only makes Anxi become the birthplace of asexual reproduction of tea trees in the country, but also allows people from all over the world to have a chance to taste Tie Guanyin in pure variety and high quality. Photographed by Li Zihong

福建安溪铁观音茶文化系统,是以传统铁观音品种选育、种植栽培、植保管理、采制工艺和茶文化为核心的农业生产系统,以及该系统在生产过程中孕育的生物多样性,发挥的生态系统功能,呈现的人文和自然景观特征。该文化系统推广了带状茶—林模式。图为安溪举源茶山风光。(李子宏 摄)

Anxi Tie Guanyin Tea Culture System is an agricultural production system with breeding, cultivation, tea protection and management, picking and processing techniques and tea culture as its core, combined with the biodiversity bred during the production process, the ecosystem functions the system played and the human and natural landscape features it presented. The cultural system promoted the belt-type tea-forest model. The picture shows the scenery of tea in Juyuan Village, Anxi County. Photographed by Li Zihong

世界人工栽培稻源头
江西万年稻作文化系统

水稻是中国人生活中一种举足轻重的存在。为了种植水稻，人们不断与自然环境做斗争，在山地上修筑梯田、在平原上挖凿水利工程……上溯到神农时代，稻被列为"五谷"之一；大禹为种植水稻而治理"九河"；《管子·地员》中记录了10个水稻品种，使中国成为世界上最早对水稻品种进行文字记录的国家，而这一切，都要从对野生稻的人工驯化说起。

我国的野生稻生长在南方平原地区的池塘、沼泽等海拔600米以下的湿地中。居住在这些地区的先民们，在采集野生稻谷为食时发现，自然落谷能够萌发水稻生长。于是，中国人开始了对野生稻的人工栽培。在江西万年县的仙人洞和吊桶环遗址被发掘之前，人们认为距今7000多年前的浙江河姆渡，是世界人工栽培水稻的发源地。1995年经中美联合农业考古发掘，认定万年为当今所知世界最早的栽培稻遗址，由此将世界稻作起源由7000年前推移到12000年前。当原始先民在仙人洞种下第一株水稻时，便注定了万年会在人类稻作历史上留下浓墨重彩的一笔。江西万年稻作文化系统即以贡米栽培为特色，并经历"野生稻—人工栽培野生稻—栽培稻"的水稻驯化过程的稻作文化系统。

江西万年县位于鄱阳湖东南岸，是典型的南方稻作区。亚热带季风气候令万年降水丰沛、热量充足。从有文字记载开始，水稻就是万年农业的主要栽培作物。万年县辖的裴梅镇，山高坳深，泉流密布，水土中含有丰富的矿物质，为水稻提供了独特的生长环境。这里的水稻品种极接近野生稻，是原始的栽培稻品种，具有抗病虫、抗逆境及不可移植性。其稻米体长粒大，品质上乘，明代时便作为贡米，被指定"代代耕作，岁岁纳贡"。这也使万年获得"贡米之乡"的称号。

稻作给万年留下了深刻的文化烙印。在万年县，"小暑小割，大暑大割"等根据节气总结的农事物候经验一直被遵守着；"敬老有福，敬土有谷""开秧门""祭谷王"依旧是从事稻作劳动的农民们心中的信仰；做米糖、酿米酒的手艺在万年人的手中代代传承。万年人种植水稻已不再单纯为了果腹，更成了一种文化信仰和农耕精神的传承纽带。

地理位置：北纬28°30′—28°54′、东经116°46′—117°15′，地处江西省上饶市万年县。
气候特点：亚热带季风性湿润气候。
认定时间：2010年被联合国粮农组织列为全球重要农业文化遗产，2013年被农业部列为首批中国重要农业文化遗产。

Geographical location: 28°30′-28°54′N, 116°46′-117°15′E; located in Wannian County, Shangrao City, Jiangxi Province.
Climate type: subtropical humid monsoon climate.
Time of identification: In 2010, it was listed as one of the GIAHS by the FAO. In 2013, it was listed as one of the first batch of the China-NIAHS by the Ministry of Agriculture.

The Manual Rice Cultivation Place in the World

Wannian Traditional Rice Culture System

Rice is a pivotal existence in Chinese life. To plant rice, people have been in competition with the natural environment for a long time. They build terraces on the mountains, and dig water conservancy projects on the plains...Back to the Shennong era, rice was one of the "five cereals"; King Yu tamed the nine branches of Yellow River for planting rice; ten rice varieties recorded in the passage Diyuan in *Guan Zi*, a compilation book of speeches of various schools in the pre-Qin period make China the first country in the world to record rice varieties. But all of these start with the artificial domestication of wild rice.

China's wild rice grows in ponds and swamps and other wetlands of 600 meters below sea level in the southern plains. The ancestors who lived in these areas found that natural grains from the wild rice could grow up. Then, the Chinese started the artificial cultivation of wild rice. Before the excavation of the Cave of Celestial Being and the "Hang Bucket Ring" Site in Wannian County, Jiangxi Province, Hemudu Site in Zhejiang Province, more than 7,000 years old, was believed the birthplace of artificially cultivated rice in the world. In 1995, after the Sino-US joint agricultural archaeological excavation, the two sites was identified as the earliest cultivated rice site in the world, and the origin time of the world rice plants were changed from 7000 years ago to 12,000 years ago. When the ancestors planted the first rice in Cave of Celestial Being, it was doomed to leave an indelible mark on the history of rice. The Wannian Traditional Rice Culture System is characterized by the cultivation of tribute rice, and has experienced the domestication process from wild rice to cultivated wild rice and then the cultivated rice.

Wannian County, located in the southeastern shore of Poyang Lake, is a typical rice-growing area of Southern China. The subtropical monsoon climate brings abundant rainfall and sufficient heat. As far back as recorded history goes, rice has been the main cultivated crop in Wannian. Peimei Township in Wannian County, with high mountains, dense nets of rivers and rich minerals in the soil, provides a unique environment for the growth of rice. The rice here is very close to wild rice, and it is a primitive cultivated rice variety with resistance to pests and hostile environment and is non-portable.

Its rice grains are large in size and high in quality. It was paid as a tribute in the Ming Dynasty and was designated as the crop that should be "cultivated from generation to generation and paid as the tribute year after year". Wannian was also earned it reputation as "The Land Flowing Tribute Rice".

Rice has left a deep cultural brand for Wannian. Some farming phenology experience based on solar terms such as "harvest slightly in Slight Heat, while harvest greatly in Great Heat"; farmers still hold the belief in their hearts that "respect the eldership will get good fortune and revere the earth will get high field"; rice-transplanting ceremony, the worship of the god of rice are still be held every year; The crafts of making rice sugar and brewing rice wine are descended from generation to generation. Wannian people plant rice no longer just for satisfying their hunger, but also for the heritage of the cultural beliefs and farming spirit.

始建于1512年的江西万年县，位于江西省东北部、鄱阳湖东南岸，历史厚重、秀美神奇、人文鼎盛，享有"世界稻作文化发源地""中国贡米之乡"的美誉。早在12000年前，这里已是天地形胜、稻花飘香。图为当今所知世界最早的栽培稻遗址——仙人洞前平原。（李鹏春 摄）

Wannian County, founded in 1512, is located in the northeastern part of Jiangxi Province and on the southeastern shore of Poyang Lake. It has a rich history, beautiful scenery and prosperous humanity culture. She enjoys the reputation of "the Birthplace of The World's Rice Culture" and the "Land Flowing Tribute Rice". Wannian County have been full of fragrance of rice flowers and richly endowed by nature for 1,2000 years. The picture shows the earliest cultivated rice ruins in the world well-known today—plain before the Cave of Celestial Being. Photographed by Li Pengchun

仙人洞位于距万年县县城 12 千米的大源盆地小荷山山脚，呈狭长形，为溶洞构造；吊桶环遗址是原始居民在仙人洞之前的栖息地、狩猎的临时性营地和屠宰场。生活在仙人洞、吊桶环的古万年人，创造了灿烂的远古文明，为万年深深地打下了农耕文明的烙印。图为今天的吊桶环和仙人洞。（吴青华 摄）

The Cave of Celestial Being is located at the foot of Xiaohe Mountain in Dayuan Basin, 12 kilometers away from Wannian County. It is in a karst cave and extends in a long and narrow way. Hang Bucket Ring Site is the shelter, temporary camp and slaughter house of ancestors. The ancient Wannian people who lived in the Cave of Celestial Being and the Hang Bucket Ring Site created a splendid ancient civilization and deeply imprinted Wannian with the brand of farming civilization. The pictures show the Hang Bucket Ring Site and the Cave of Celestial Being. Photographed by Wu Qinghua

华东地区·East China/江西省·Jiangxi Province

万年仙人洞主洞空旷幽深，长60米，宽25米，高3米，可容纳1000余人。左右各有支洞，深长莫测，是我国首次发现的从旧石器时代向新石器时代过渡的人类活动文化遗迹。图为今天的仙人洞内景，里面有原始居民生活场景的雕塑。
（李鹏春 摄）

The main cave of the Cave of Celestial Being is vast and deep. It is 60 meters long, 25 meters wide and 3 meters high and can accommodate more than 1,000 people. Branch holes with unpredictable length are on the left and right. The cave is the first cultural site of human activities in transitional stage from the Paleolithic Age to the Neolithic Age found in China. The picture shows the interior of the cave. There are some sculptures of the primitive residents' living scenes. Photographed by Li Pengchun

万年贡谷原名"坞源早",接近野生稻形态特征,是古人不断从生产实践中逐渐选育而成,是带有显著野生稻特性的原始栽培稻品种,体长粒大、形状如梭、其白如玉、光洁透亮。图为万年贡谷长长的谷芒。(吴青华 摄)

The original name of Wannian Tribute Rice was "Wuyuan Zao". Rice growing there is close to the morphological characteristics of wild rice. The rice, a cultivar that is gradually bred from the panting practice, has long grains which are shaped like as shuttles, smooth and translucent as jades. The picture shows the long awns of the rice growing from Wannian fields. Photographed by Wu Qinghua

在数千年的农耕文化习俗中，万年人民总结出一套从良种培育更新、播种移栽、田间管理、收割贮存到精制加工等的传统生产方式。早稻与晚稻栽种有所区别，"早稻顺手栽，晚禾栽到隔"，就是说早禾栽种轻轻用力，而晚禾栽种用力到位。图为当地农民在拔秧苗。（吴青华 摄）

In the thousands of years of farming activities, Wannian people have summed up a set of traditional production methods, such as breeding upgradation, seed-sowing and transplanting, field management, harvesting and storage, and refining and processing. There are little differences between early season rice and late rice. The early season rice should be planted with gentle strength while the late rice should be paid much more appropriate strength. The picture shows local farmers are pulling out the seedlings. Photographed by Wu Qinghua

万年作为稻作文化起源地，沿袭传承着一整套耕作技术与传统习俗，作为五谷之首的"稻谷"有着独特的耕作方式与技巧，全县许多乡村仍保留着这一古老的耕作传统。图为贡米产地和传统牛耕。（吴青华 摄）

As the birthplace of rice culture, Wannian inherits a whole set of farming techniques and traditional customs. Rice, the "elder brother" of the "five cereals" has its unique farming methods. This ancient farming tradition is still preserved in many villages in the county. The pictures show the tribute rice field and traditional cattle farming. Photographed by Wu Qinghua

如今，万年很多地方还保留着"敬老有福，敬土有谷""开秧门""祭谷王"等农耕信仰，这些信仰在维系农耕社会秩序、净化人们心灵、保护自然环境等方面都发挥了重要作用。与之相关发展出很多民间艺术，如民歌、灯彩、戏剧、手工艺等，其中太平跳脚龙灯，又名"矮脚龙灯"，已有1000多年历史。（吴青华 摄）

Rice growers here still hold the belief in their hearts that respect the eldership will get good fortune and revere the earth will get high field; rice-transplanting ceremony, the worship of the god of rice are still be held every year. These beliefs and cultural customs play important roles in maintaining the order of farming society, purifying people's minds, and protecting the natural environment. Many folk arts are related to the rice, such as folk songs, lanterns, drama crafts, etc. Jumping Dragon lanterns, also known as short dragon lanterns enjoy a history of more than 1,000 years. Photographed by Wu Qinghua

最大的客家梯田
江西崇义客家梯田系统

在中国的 56 个民族之外，还有着一个特殊的族群。他们不是一个单一的少数民族，却有着一个固定的称谓——客家人。他们经历了战乱或饥荒，从家乡迁居于别处，土著居民以"客家人"来称呼他们。

明嘉靖年间（1522—1566 年），倭寇在东南沿海地区活动猖獗，烧杀抢掠使当地居民广受侵害。饱受战乱之苦的人们，决定迁居。崇义县位于江西省的西南部，是赣、粤、湘三省交界之地，山地和高丘占到了崇义县域土地的 90%。这里山连着山，岭叠着岭，山脉重峦叠嶂，群峰连绵起伏，外贼不易侵入，是较为理想的安全之地。避祸乱的人们走到这，便决定定居下来。为了维持生计，客家先民在岭上依山建房，开山凿田。崇义客家梯田即以客家文化为特色的稻作梯田文化系统。梯田始建于元朝，完工于清初，距今已有 800 多年的历史。明代理学家王守仁撰写的《立崇义县治疏》介绍了从广东返迁此地的客家先民胼手胝足开垦梯田的业绩。

彼时的崇义刚刚立县（崇义县建于公元 1517 年），"九分山半分田，半分道路、水面和庄园"的地理条件令这里人烟稀少，农业水平落后。当地土著居民从元代时开发的几分梯田，零星散落在群峰之间，只能满足当地人的正常生活需要。突然涌入的人口，使得田地一下子紧张起来。迁到此的客家人迫于生计，开始依山建房、开山造田。经历过战乱的客家人更加珍惜和平的时光，他们勤劳、勇敢、朴实。到清初，崇义县上堡乡的梯田已经形成规模。这片梯田面积达 4 万亩，层层盘绕在海拔 2000 多米的赣南第一高峰齐云山山脉之中。在这片中国最大的客家梯田中，位置最高的田块海拔达 1260 米，最低的则位于海拔 280 米处。层级最多的梯田有 62 梯层，其中大多数是只能种一行、两行禾稻的"带子丘"和"青蛙一跳三块田"的碎田块。

历经百年的共同开拓和劳作，当初迁居于崇义的客家人，早已与原住民一起成为这片土地的主人。他们有了自己的农耕谚语、田埂文化和饮食文化。由客家人带来的"舞春牛"祈丰收、平安的习俗，在经历了与当地文化的结合和演变后，也已成为崇义在节日中不可或缺的传统节目。

地理位置：北纬 25°24′—25°55′、东经 113°55′—114°38′，地处江西省赣州市崇义县。
气候特点：亚热带季风湿润气候。
认定时间：2014 年被农业部列为第二批中国重要农业文化遗产，作为"中国南方山地稻作梯田系统"之一于 2018 年被联合国粮农组织列为全球重要农业文化遗产。

Geographical location: 25°24′-25°55′N, 113°55′-114°38′E; located in Chongyi County, Ganzhou City, Jiangxi Province.
Climate type: subtropical humid monsoon climate.
Time of identification: In 2014, it was listed as one of the second batch of the China-NIAHS by the Ministry of Agriculture. In 2018, as one of the Rice Terraces in Southern Mountainous and Hilly Areas in China, it was listed as one of the GIAHS by the FAO.

The Largest Hakka Terraces
Chongyi Hakka Terraces

In addition to the 56 ethnic groups in China, there is a special ethnic group. They don't belong to any minority, but they do have a name of their own—Hakka. They suffered from war and famine and then moved elsewhere from their hometowns. The aboriginal call them "Hakkas".

During the years reigned by Emperor Jiajing in the Ming Dynasty, Japanese pirates swept the southeastern coastal areas, they slaughtered innocent victims, burned countless buildings, and locals suffered a lot and determined to run out of the hell. Chongyi County is located in the southwestern part of Jiangxi Province and is the junction of the three provinces of Jiangxi, Guangdong and Hunan. Mountains and high hills account for 90% of the land. With undulating, rolling and puzzling mountain roads, this was an ideal to hide from the war. People looking for a peaceful land decided to settle down here. In order to maintain their livelihood, the ancestors of Hakka built houses, reclaimed land and started their lives on the hills.

Chongyi Hakka Terraces is a rice terraced culture system featuring Hakka culture. These terraces were first built in the Yuan Dynasty and completed in the early Qing Dynasty. It has a history of more than 800 years. Wang Shouren, a Neo-Confucianist in the Ming Dynasty introduced that Hakka ancestors reclaimed the land with perspiration in the letter to the throne about his suggestion of the establishment of Chongyi County.

At the very beginning of establishment of Chongyi County in 1517, "nine tenths of the earth were cover by mountains and the rest were fields, paths, waters and manor." People there were sparsely populated and the agricultural level was backward. The terraces developed by the native from the Yuan Dynasty, scattered between the peaks, could only meet the needs of the local people. Those Hakka had to build houses and reclaim on mountains because of the shortage of field caused by the sudden influx of people.

Hakka people who had experienced wars cherished the time of peace. They were hardworking, courageous and simple. Terraces in Shangpu Township, Chongyi County is on a large scale. It covers an area of 40,000 mu, and is layered on the Qiyun Mountain Range, the highest peak of southern Jiangxi, at an altitude of over 2,000 meters. Among the largest Hakka terraces in China, the highest terrace bar is 1,260 meters above sea level, while the lowest is at 280 meters above sea level. The terraces with the most stairs have 62 stairs, most of which are "belt-shaped" fields that can only grow one row or two rows of rice or even can be jumped over by a frog.

After a hundred years of hardworking, the Hakka who migrated to Chongyi had already become the co-owner of this land together with the native people. They have developed their own farming slang, terrace dykes culture and food culture. "Spring Cattle Dance" praying for good harvest and peace, has become an indispensable traditional program in Chongyi after experiencing the combination and evolution with local culture.

崇义客家梯田位于江西省崇义县上堡、丰州、思顺三个乡，坐落在海拔2000多米的赣南第一高峰齐云山山脉之中，是中国南方山地稻作梯田的组成部分。这里森林覆盖率高，生物多样性丰富，有着完善的自然生态系统。（曾正东 摄）

Located in the three townships of Shangpu, Fengzhou and Sishun in Chongyi County, and seated at Qiyun Mountain Range, Jiangxi Province, Chongyi Hakka Terraces is located in the Qiyun Mountain Range, the highest peak of southern Jiangxi Province at an altitude of over 2,000 meters. Chongyi Hakka Terraces are part of the rice terraces in southern China. It has high forest coverage, rich biodiversity and perfect ecosystem. Photographed by Zeng Zhengdong

千百年来，客家人采用原生态的耕作技术，变自然生态为农业生态，属于原生态保护较好的农区。和很多地方的梯田耕作模式一样，由于梯田独特的地理特点，这里传统的耕作技术延续至今。图为春耕和当地农民在拔秧。（曾正东 摄）

For hundreds of years, Hakka people have adopted the original ecological farming techniques to change the natural ecology to agro-ecology, which set a good example in terms of environment protection. Like the terraced farming model in many places, the traditional farming techniques here continue to this day because of the unique geographical features of the terraces. The pictures show the spring ploughing and the local farmers in the transplanting. Photographed by Zeng Zhengdong

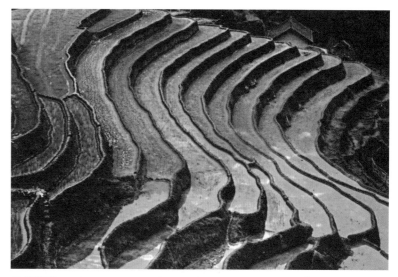

崇义客家梯田因主要分布于上堡乡，故常被称为上堡梯田。梯田依山势开建，连绵数百亩，又有零星村落点缀其间。在耕作期里，泉水自山顶向山下逐层灌溉，气象万千；收获期里，五颜六色的农作物又给梯田增添了无限生机。图为春、夏、秋不同季节的崇义客家梯田景观。（曾正东 摄）

Chongyi Hakka Terraces are also called as Shangpu terraces because they are mainly distributed in Shangpu Township. The hundreds of terraces are built on the mountain, with and sporadic villages dotted around. During the cultivation period, the spring water is irrigated from the top to the bottom of the mountain. The colorful crops in the harvest period add infinite vitality to the terraces. These pictures are the landscape of Chongyi Hakka Terraces in spring, summer and autumn. Photographed by Zeng Zhengdong

一块块、一排排、一垄垄田犹如横在天地间的一部厚重史诗，写满了一代代客家人的智慧和汗水，成为客家农耕文明的一道奇观。（曾正东 摄）

These terraces bars are like a thick epic written between the sky and the earth. They are filled with the wisdom and sweat of generations of Hakka people, and become a spectacle of Hakka farming civilization. Photographed by Zeng Zhengdong

客家先民发展出了丰富多彩的与梯田开垦和劳作相关的民间文化。其中最具代表性的是"舞春牛"。在客家人的心目中,千百年来和他们一道辛勤耕耘这片土地的牛就是"神"。图为当地人在表演"舞春牛"。(曾正东 摄)

Hakka ancestors created rich and colorful folk cultures related to terraces. The best representative of them is the "Spring Cattle Dance". The cattle that have worked hard with them for thousands of years are their god. The picture shows the dance. Photographed by Zeng Zhengdong

沙地生态治理与经济发展的典范
山东夏津黄河故道古桑树群

桑树在中国是一种常见的林木，它承载了中国人很多美好的情感：可以表达对心上人的喜爱，也能表示对父母双亲的敬爱。种植桑树的"性价比"很高：其叶可养蚕，其果可果腹，其树干是为良材。据考古发现，中国自新石器时期便开始种植桑树。先秦时期，桑树种植已经较为普遍。从《诗经》中频繁出现的以"桑"为意象的诗歌中，可以看出当时黄河中下游地区已经开始桑蚕丝织活动。山东是种桑养蚕的先行区，《管子》《左传》等古籍中多有提到齐鲁两国（今山东地区）的桑蚕故事。

夏津县位于山东西北部，历史上黄河干流两次流经于此。公元前602年，黄河第一次大改道，呈西南—东北向流经夏津；公元11年，黄河再次大改道，此段河流被废弃；公元1048年，黄河又一次大改道，再次流经夏津；公元1194年，黄河继续大改道，此河废弃。黄河干流两次流经夏津、多次改道和泛滥大决口，给夏津留下了30万亩沙丘地，也给当地居民带来了无数烦恼。在夏津，流传着这样一首歌谣："无风三尺土，有风沙满天，关门盖着锅，土饭一起咽。"每当风沙来临时，这首歌谣就是夏津人民生活的真实写照。元代时，夏津人民开始尝试大面积种树，造林固沙。桑树生命力旺盛，耐旱耐贫瘠，对土壤的适应性极强。它的根系发达，根幅是树冠的几倍甚至几十倍，是固沙的能手。此外，桑树兼具经济效益，成了为夏津黄河故道固沙的不二之选。

造林固沙的信念在夏津人的心中传承了一代又一代，至清末民初，夏津黄河故道的桑树群面积达到8万亩，相传"援木可攀行二十余里"。种植桑树不仅改善了夏津人们的居住环境，还给他们的生活带来了生机。明清时期，桑蚕丝织业一直是夏津人的主要劳动收入；从不迟到的桑葚，在饥荒之年，填饱了夏津人的肚子。如今，夏津黄河故道遗存的核心区6000亩桑树，也成为世界上保存最完整的古桑树群。其中千年以上古桑树百余株，百年以上古桑树20000多株，涉及今天的12个村庄，被命名为"中国葚果之乡"，是远近闻名的"中国北方落叶果树博物馆"。历经时间的洗礼，桑树形态各异，但依然枝繁叶茂，散发着蓬勃的生命活力。

地理位置：北纬36°53′—37°10′、东经115°45′—116°16′，地处山东省德州市夏津县。
气候特点：温带季风气候。
认定时间：2014年被农业部列为第二批中国重要农业文化遗产，2018年被联合国粮农组织列为全球重要农业文化遗产。

Geographical location: 36°53′-37°10′N, 115°45′-116°16′E, located in Xiajin County, Dezhou City, Shandong Province.
Climate type: monsoon climate of medium latitudes.
Time of identification: In 2014, it was listed as one of the second batch of the China-NIAHS by the Ministry of Agriculture. In 2018, it was listed as one of the GIAHS by the FAO.

A Representative of Ecological Management and Economic Development in Sandy Land
Traditional Mulberry System in Xiajin's Ancient Yellow River Course

Mulberry is a common variety in China, it carries many beautiful emotions of the Chinese. It can express the love for your sweetheart or the respect for your parents. The "cost-effectiveness" of planting mulberry trees is very high: its leaves can raise silkworms, its fruit can satisfy your stomach, and its trunk is a good timber. According to archaeological findings, China began planting mulberry trees during the Neolithic period. Its cultivation was more common in the pre-Qin period. From the poems often appearing in *The Book of Songs* with the image of "Mulberry", it can be seen that the silkworm weaving activities had begun in the middle and lower reaches of the Yellow River. Shandong Province is the pioneering area for mulberry-planting and sericulture. The ancient books such as *Guan Zi* and *Zuo' Commentary* often mention the silkworm stories of Qi and Lu countries (now Shandong Province).

Xiajin County is located in the northwest of Shandong Province. The Yellow River flowed through it twice in the history. In 602 BC, the Yellow River rerouted for the first time, flowing southwest-northeast through Xiajin; in 11AD, the Yellow River was once again diverted, and the branch was abandoned; in 1048, Yellow River changed passed through Xiajin again; in 1194, Yellow River continued to be diverted, and the river was abandoned. The main stream flowed through Xiajin twice, diverted many and overflowed many times. It not only left 300,000 mu of sand dunes for Xiajin, but also brought countless troubles to local residents. In Xiajin, a ballad goes like that: three feet of sands in windless day, a skyful of sands in windy day; don't bother to close the door and lid the pot, I will eat the rice with the sand. It is a true portrayal of Xiajin people's life whenever the sandstorm comes.

Xiajin people began to try to plant large areas of trees for sand fixation during the Yuan Dynasty. Mulberry has strong vitality, drought and infertility tolerance and is highly adaptable to the soil. Its root system is high developed, and its root width is even several times wider than that of the crown. It is a master of sand fixation. In addition, the mulberry tree also has economic benefits, so it has become the best choice for sand fixation in the Xiajin Yellow River.

The belief in afforestation and sand fixation has been passed down from generation to generation in the hearts of Xiajin people. By the end of the Qing Dynasty and the early Republic of China, the area of the mulberry trees in the Ancient Yellow River Course reached 80,000 mu. it was said that "contiguous branches stretches twenty miles." Planting mulberry not only improved the living environment of Xiajin people, but also brought vitality to their lives. The silkworm weaving industry was always the main income resource of Xiajin people during the Ming and Qing Dynasties; the mulberry filled the Xiajin people's stomach in the years of famine. Today, the 6,000 mu of mulberry trees in the core area of Ancient Yellow River Course remains the best preserved ancient mulberry trees in the world. Among them, there are more than 100 ancient mulberry trees over a thousand years old, and more than 20,000 ancient mulberry trees over 100 years old. Twelve villages planting the ancient trees today are named "China's Hometown of Mulberry" and are known as the "Museum of Deciduous Fruit Trees North China ". After the baptism of time, though in different poses, they are still flourishing and exuding vitality.

中国树龄最高、规模最大的古桑树群位于山东省德州市夏津县东北部黄河故道中。那里是黄河曾流经的夏津腹地，如今遗存一片6000亩的古桑树群，是目前世界上保存最完整的古桑群。图为位于今天夏津颐寿园内的著名的卧龙桑与腾龙桑，两棵树一高一矮，一壮一弱，形成"双龙争霸"的古桑景观，合称为"双龙树"，树龄在1000年以上。（马志勇 摄）

The oldest and largest ancient mulberry trees group in China is located in the Ancient Yellow River Course in the northeast of Xiajin County, Dezhou City, Shandong Province. It is the hinterland of Xiajin where the Yellow River once flowed. Today, there is 6,000 mu of ancient mulberry trees, the best preserved ancient mulberry group in the world. The picture shows the famous Wolong and Tenglong mulberry trees in Longevity Park today. One is tall and strong and the other is short and small, showing a hegemony between them. They are known as the double dragon trees of 1,000 years old. Photographed by Ma Zhiyong

山东夏津古桑树种植时期跨元、明、清三朝，这片古桑树群是夏津世世代代老百姓防风治沙的见证。图为夏津县黄河故道森林公园的最南端，这里有一片千亩古桑园，当地人因久食葚果，年寿逾增，故桑树又称"颐寿树"，这片园子就叫"颐寿园"。园区面积近 1000 亩，100 年以上的古葚树 3000 多株，1000 年以上的 100 余株。（马志勇 摄）

Cultivated in the dynasties of Yuan, Ming and Qing, these ancient mulberry trees witness how Xiajin's descendant to prevent wind and sand. The picture shows the southernmost end of the forest park of the Ancient Yellow River Course in Xiajin County. There is an ancient mulberry trees park of a thousand mu. The local people have eaten mulberries a lot and their annual lifespan is more than ever. Therefore, the mulberry tree is also called "longevity trees" and the park is called Longevity Park. The park covers an area of nearly 1,000 mu, with more than 3,000 ancient mulberry trees over 100 years and more than 100 plants over the millennium. Photographed by Ma Zhiyong

种植桑树，防风固沙的同时，夏津县的劳动人民还探索出了一套桑树"种植经"。他们用土炕坯围树、畜肥穴施、犁伐晒土等方法施肥和管理土壤，用油渣刷或塑料薄膜缠树干的方法防治害虫，天然无公害。（闵庆文 摄）

Xiajin people also summarize a set of mulberry lection while planting mulberry trees for wind prevention and sand fixation. They use sands to surround the tree, apply manure in small holes near the plants, plow and then dry the soil in the sun; they also use oil residue or plastic film for pest control, which is pollution-free. Photographed by Min Qingwen

华东地区・East China/ 山东省・Shandong Province

当地流传着"打枣晃葚"的说法，即采用"抻包晃枝法"采收桑葚。每年"小满"前后，正是葚果成熟时，是采摘、品尝葚果的最好季节。冠盖巨大的树枝上，累累果实如繁星缀空，村人在树下抻起长单，持长杆晃动树枝，收获葚果。（马志勇 摄）

Mulberries are harvest by shaking them off and collected with a big piece of cloth underneath the tree. The picture shows the best season for harvesting and tasting when the mulberries are ripe in Grain Full(8th solar term). Mulberries crowds like a skyful of stars on the huge crown. Growers stretch big cloth underneath the tree, shake the branches with long poles and harvest the fruits. Photographed by Ma Zhiyong

古桑树群群落结构复杂、生态稳定。群落以桑树为主，间有其他落叶乔木、灌木和草本。数百年的古桑，枝繁叶茂，根系发达，冠幅10米的古桑树，年产葚果400千克，鲜叶225千克。夏津的桑树经过多年改良，如今的葚果颗粒饱满、果肉肥厚，汁溢鲜嫩，味甘如蜜，状似草莓，不仅个头大，而且灌浆后口感甜蜜。（马志勇 摄）

The structure of ancient mulberry tree community is complex and ecologically stable. The community is dominated by mulberry trees, scattered with deciduous arbors, shrubs and vegetation. The ancient mulberry trees of hundreds of years have leafy branches and high developed roots. An ancient mulberry tree with a crown of 10 meters can be harvested with an annual output of 400 kilograms of mulberries and 225 kilograms of fresh leaves. The mulberry tree of Xiajin has been improved for many years. Today, the mulberries growing there are full, thick, juicy and fresh. They are tasted like honey and looked like a plump strawberry. Photographed by Ma Zhiyong

近年来，夏津县在葚果产业发展上，进行了深入研究，探索出了一条葚果产品深加工的新路子。新近建立的椹产品研究所，目前已经成功研制和生产了高附加值的葚果酒、葚果醋、桑叶茶、桑葚干等一系列旅游产品。图为以葚果为原料进行葚果酒的加工制作。（周坤 摄）

In recent years, Xajin county conducted deep research on the development of fruit industry, and explored a new way of deep processing of fruit products. The research institute of mulberry products has been established, and a series of tourism products such as high value-added mulberry wine, fruit vinegar, mulberry tea and dried mulberries have been successfully developed and produced. The pictures show the processing of mulberries wine. Photographed by Zhou Kun

桑黄是一种真菌，因寄生于桑树而得名，是一种名贵中药。在中国，桑黄的使用从汉朝起至今已经有 2000 多年的历史。《本草纲目》记载，桑黄能利五脏，宣肠胃气，排毒气；现代研究证实，桑黄多糖，能够缓解疼痛、食欲不振、体重减轻及疲劳倦怠等癌症特有的症状，提高生活品质。图为长在夏津古桑树上的桑黄。（董玉龙 摄）

Phellinus igniarius is a fungus named for parasitizing on mulberry trees. It is a valuable Chinese medicine. It has been in use for more than 2,000 years since the Han Dynasty. *Compendium of Materia Medica* records that it can help exhaust the poisonous gas produced by intestines and stomach and is good for the five internal organs. Modern research has confirmed that phellinus igniarius can relieve the symptoms of cancer, such as pain, loss of appetite and weight, and fatigue, so it can help improve the quality of life. The picture shows the phellinus igniarius growing on the mulberry tree. Photographed by Dong Yulong

华东地区·East China/ 山东省·Shandong Province

夏津黄河故道古桑树形态各异，或群或孤，给人一种历史沧桑之感。有的古桑树经过数次雷击却依然枝繁叶茂，有的古桑树承载着一段段美丽动人的古老的故事。图为夏津黄河故道森林公园内树龄最长的葚树，树龄已有1500多年，被称为"葚树王"，虽然经历了千年的风雨沧桑，但仍枝繁叶茂，体现出坚韧的生命力。（马志勇 摄）

The ancient mulberry trees in Ancient Yellow River Course gather in a group or stand alone, giving people a sense of historical vicissitudes. Some are still leafy after several lightning strikes, and some carry beautiful and touching stories. The picture shows the oldest mulberry tree in the forest park. It is more than 1,500 years old and is known as the "King of Mulberry Tree". Although it has experienced thousands of years of vicissitudes, it is still flourishing and showing its tough vitality. Photographed by Ma Zhiyong

华中地区·Central China

湖北省

砖茶之源——湖北羊楼洞砖茶文化系统

湖南省

南方稻作文化与苗瑶山地渔猎文化融合体系——湖南新化紫鹊界梯田

世界原始稻作文化的"活化石"——湖南新晃侗藏红米种植系统

砖茶之源
湖北羊楼洞砖茶文化系统

18世纪开始,一个名不见经传的小镇"羊楼洞"开始成为全国贸易的焦点。它可媲美当时的经济重镇汉口,被人们称作"小汉口"。从羊楼洞走出的羊楼洞砖茶,驰名海内外,在俄罗斯等地被认为是中国好茶的代表。

羊楼洞镇位于湖北蒲圻(今赤壁市)西南,是湘、鄂两省的交界之处,属幕埠山脉北麓余峰、湘鄂交界的低山丘陵地带,是茶马古道的三大源头之一。这里从汉晋时期开始种茶。唐代以后,茶叶种植全面铺开,蒲圻地区"唯以植茶为业",茶叶成为羊楼洞的主要产业。湖北赤壁羊楼洞砖茶文化系统,就是以茶叶种植和青砖茶加工为特色的茶文化系统。

明末清初之际,海上丝绸之路因中国的闭关锁国政策逐渐消沉,中俄万里茶道的另一条路线福建武夷山茶路因战乱中断,中国茶叶的外销通道,向以湖北茶区为中心的路线倾斜。到19世纪,茶叶重镇羊楼洞发展成为中俄茶叶国际商道的起点。这条贸易通道以羊楼洞为起点,经汉口,在水路、陆路多次转运后,到达中俄边境的恰克图,再由此向西。青砖茶就是这条贸易通道北向运送的主要物资。

青砖茶又称"砖茶",属于黑茶类,因外形像砖而得名。黑茶特指颜色为褐色的后发酵茶类,唐朝就有"渠江有薄片,一斤八十枚,其色如铁"之说,那是因为古时交通不便,运输困难,必须减少货物体积,于是将茶蒸压成团块。在加工成团的过程中,要经过湿坯堆积,所以毛茶的色泽逐渐由绿变黑,成品团块茶叶的色泽为黑褐色,并形成了茶品的独特风味,这就是黑茶的由来。

宋景德年间(1004—1007年),两湖的茶叶曾被官府销往西域及蒙古地区,进行茶马贸易。当时为便于运输,人们将茶叶用米浆粘合成饼状,制成茶饼。到了明永乐年间(1403—1424年),为了减少运输途中的损耗、降低运费成本,羊楼洞的制茶大户将青茶茶叶制成圆柱形的"帽盒茶"。后来,茶叶需要运送到更远的西伯利亚,于是又被制成了砖块形的"砖茶",这是在当时的生产条件下,能够最大幅度减少运输损耗的制茶工艺。

地理位置:北纬29°28′—29°59′、东经113°32′—114°13′,地处湖北省咸宁市赤壁市。
气候特点:亚热带季风气候。
认定时间:2014年被农业部列为第二批中国重要农业文化遗产。

Geographical location: 29°28′-29°59′N, 113°32′-114°13′E; located in Chibi City, Xianning City, Hubei Province.
Climate type: subtropical monsoon climate.
Time of identification: In 2014, it was listed as one of the second batch of the China-NIAHS by the Ministry of Agriculture.

The Origin of Brick Tea

Yangloudong Brick Tea Culture System

Beginning in the 18th century, a little-known town "Yangloudong" began to become the focus of national trade. It is comparable to the economic center of Hankou at that time, and it is called "Little Hankou." Yangloudong Brick Tea from Yangloudong, is well-known at home and abroad and is regarded as the representative of Chinese good tea in Russia and other places.

Yangloudong Town is located in the southwest of Puqi (now Chibi City), Hubei Province, at the junction of Hunan and Hubei provinces. It belongs to the northern part of the Mufu Mountain Range and a low hilly area bordering on Hunan and Hubei. It is one of the three major sources of the tea-horse roads. Here, tea was grown from the Han and Jin Dynasties. After the Tang Dynasty, tea planting was spread out in an all-round way. Tea planting was the only industry in Puqi area. Tea became the main industry of Yangloudong. The brick tea culture system of the Chibi Yangloudong in Hubei Province is a tea culture system featuring tea planting and green brick tea processing.

At the end of Ming Dynasty and the beginning of Qing Dynasty, the Maritime Silk Road was gradually subsided due to China's policy of closure and blockade. Another route of the Sino-Russo Tea Road of Ten Thousand Miles was interrupted by the war in Wuyi Mountain, Fujian Province, and the export channel of Chinese tea tilted to the tea production area of Hubei.

In the 19th century, Yangloudong as a major tea town became the starting point of the Sino-Russo international tea business. Tea products were transported from here via Hankou, and transferred till Kyakhta on the Sino-Russo border after many times on the water and land, and then went west on. Green brick tea is the main material for this trade route to the north.

Green brick tea, also known as brick tea, falls into dark tea and is named for its brick-like shape. Dark tea refers to the post-fermented tea with a brown color. In the Tang Dynasty, there was the saying that there were thin slices of Qujiang, 80 pieces for half kilo, and its color was like iron. It was because the transportation then was inconvenient that the tea leaves were then steamed and pressed into a solid form. In the process the tea leaves were piled up into wet billets, so the color of the raw tea leaves gradually changed from green to black. As a result, the color of the finished tea leaves was dark brown, and flavor of the tea was unique, which is the origin of dark tea.

During the Jingde Period of the Northern Song Dynasty, the tea leaves of Hubei and Hunan provinces were sold to the Western Region and Mongolia by the government. At that time, for the sake of transportation, the tea was glued into a cake with rice syrup to make a tea cake. In the Yongle period of Ming Dynasty, in order to reduce the loss during transportation and the cost of freight, the tea-making households in Yangloudong made green tea leaves into cylindrical "hat-box tea". Later, the tea was needed to be further transported to Siberia, and it was made into a brick-shaped "brick tea", which was the tea making process that could minimize the losses in transportation under the conditions at that time.

赤壁地区大部分属微酸性黄红壤土,年平均气温 13—18℃,年平均降雨量 750—1500 毫米,无霜期达 220—300 天。这里气候温和,雨量充沛,有发展茶叶生产的良好条件。图为赤壁羊楼洞如今的茶园。(夏正锋 摄)

Most of the Chibi area are slightly acidic yellow-red loam, with an average annual temperature of 13-18 ℃, an average annual rainfall of 750-1500 mm, and a frost-free period of 220-300 days. The climate here is mild, rainfall is abundant, so there are good conditions for tea production. The picture shows the tea garden of Yangloudong in Chibi today. Photographed by Xia Zhengfeng

羊楼洞砖茶在国际贸易史上展示过骄人的辉煌,在国内西域各民族的交往中,为促进民族团结起了非常重要的作用。在具有1000多年历史的茶马古道上,形成了独特的"羊楼洞砖茶文化",也自此开创了一条举世闻名的欧亚万里茶道。图为羊楼洞的欧亚万里茶道铭碑。(夏正锋 摄)

Yangloudong Brick Tea has shown great brilliance in the history of international trade and played a very important role in promoting national unity in the exchanges between different ethnic groups in the western regions of China. On the Ancient Tea Horse Road of more than 1,000 years old, a unique "Yangloudong Brick Tea Culture" was formed, and a world-famous Eurasian Tea Road of Ten Thousand Miles was created. The picture shows the inscription on the Eurasian Tea Road of Ten Thousand Miles in Yangloudong. Photographed by Xia Zhengfeng

砖茶又称"青砖茶""蒸压茶",顾名思义,就是外形像砖一样的茶叶,属黑茶种类,历史悠久。清朝乾隆年间(1736—1795年)"三玉川"和"巨盛川"两茶庄在其生产的砖茶上特别压制了代表羊楼洞三道泉水的"川"字为产品标牌,在草原牧民中信誉良好。图为"川"字品牌砖茶。(李鹏 摄)

Brick tea, or steamed tea (commonly known as green brick tea), as the name suggests, is like a brick, a part of the dark tea species, has a long history. During the Qianlong period of the Qing Dynasty, the two tea houses "Sanyuchuan" and "Jushengchuan" specially imprinted the "Chuan" character representing the three springs of Yangloudong as the product sign on the brick tea, and enjoyed a high reputation among the grassland herdsmen. The picture shows the brick tea of "Chuan" brand. Photographed by Li Peng

羊楼洞古镇，距赤壁市区 26 千米，是中俄万里茶道源头和驰名中外的"洞茶"故乡，因周边万山如羊、街市茶铺楼馆林立、泉洞众多而得名。羊楼洞系赤壁市六大古镇之一，为"松峰茶"原产地，素有"砖茶之乡"的美称。图为当地采茶女在采茶。（夏正锋 摄）

Yangloudong Ancient Town, 26 kilometers away from the city of Chibi, is the source of the Sino-Russo tea road of ten thousand miles. It is well known as the hometown of "Dong Tea" in China and abroad. It is named after the numerous surrounding mountains, teahouses in the market and the spring caves. Yangloudong is one of the six ancient towns in Chibi City. It is the origin of "Songfeng Tea" and is known as the "Hometown of Brick Tea". The picture shows the local women picking tea leaves. Photographed by Xia Zhengfeng

黑茶是利用菌发酵的方式制成的一种茶叶。黑茶的加工工艺复杂，基本工艺流程是杀青—揉捻—渥堆—干燥。黑茶因产区和工艺上的差别，主要有湖北青砖茶、云南普洱茶、湖南黑茶、广西六堡茶、四川藏茶和安徽六安黑茶等品种。图为现今赵李桥茶厂（原羊楼洞砖茶厂）工人在加工砖茶。（夏正锋 摄）

Dark tea is a kind of tea made by fermentation of bacteria. The processing techniques of dark tea are complicated, and the basic work flow is fixation-rolling-piling-drying. Due to the differences in production areas and techniques, dark tea mainly includes Hubei green brick Tea, Yunnan Pu'er Tea, Hunan Dark Tea, Guangxi Liupu Tea, Sichuan Tibetan Tea and Anhui Lu'an Dark Tea. The pictures show that workers in today's Zhao Liqiao Tea Factory (formerly Yangloudong Brick Tea Factory) are processing brick tea. Photographed by Xia Zhengfeng

《湖北茶史简述》记载，至清末，羊楼洞所产的茶叶品种有：物华、桦华、精华、月华、春华、天华、天专馨、夺魁、赛春、一品、谷芽、谷蕊、仙掌、如栀、永芳、宝蕙、二五、龙须、凤尾、奇峰、乌龙、华宝、惠兰等20余种。图为当地举行的以茶论道活动场景。（夏正锋 摄）

According to *A Brief Introduction of Tea History of Hubei Province*, by the end of the Qing Dynasty, there were more than 20 kinds of tea produced by Yangloudong, including Wuhua, Huahua, Essence, Yuehua, Chunhua, Tianhua, Tianzhuxin, Duokui, Saichun, Yipin, Guya, Gurui, Xianzhang, Ruzhi, Yongfang, Baohui, Erwu, Longxu, Fengwei, Qifeng, Oolong, Huabao and Huilan. The pictures show the scene of a tea ceremony held locally. Photographed by Xia Zhengfeng

羊楼洞曾经的辉煌留在了人们的记忆深处，由于多种因素影响，砖茶独特的制作工艺和砖茶文化就像濒危物种一样亟须得到保护和继承。砖茶是祖先留给赤壁人的财富。图为羊楼洞古镇明清石板街。（李鹏 摄）

The glory of Yangloudong has remained in people's memory. Due to various factors, the unique craftsmanship of brick tea and brick tea culture are just like endangered species and need to be protected and inherited. Brick tea is the wealth left by ancestors to the people of Chibi. The picture shows the stone plate street of Ming and Qing Dynasties in Yangloudong Town. Photographed by Li Peng

南方稻作文化与苗瑶山地渔猎文化融合体系
湖南新化紫鹊界梯田

在南方山地丘陵地区遍布的梯田中，紫鹊界梯田凭借其独特的灌溉系统脱颖而出、独树一帜。紫鹊界梯田位于湖南省中部偏西的娄底市新化县，层层叠叠的梯田布满了雪峰山余脉奉家山的大小山坡。

自秦汉时期起，居住在奉家山一带的农民就开始修筑梯田。当时，苗瑶族人为躲避其他强族的欺凌，躲进奉家山的深山里。为了生计，他们开荒拓土、修筑梯田。到了宋明时期，紫鹊界、奉家山一带已汇聚了苗、瑶、侗、汉等多个民族的居民，人口数量众多，土地供不应求，梯田面积开始快速扩张。

即使在梯田面积快速扩张时期，人们也不是毫无原则地开山造田。他们将山顶的植被保留下来，在半山腰修筑梯田，而山脚的平地则被开垦成连片的农田，形成了"山顶戴帽子、山腰围带子、山脚穿裙子"的农业生态景观。这种布局方法给紫鹊界梯田带来了"天下大乱，此地无忧；天下大旱，此地有收"的保障。

紫鹊界梯田位于亚热带季风气候区。因地形等多种原因，该地区降雨时空分布不均，旱灾常常光顾。遇旱年，山脚的稻田颗粒无收，山腰梯田里的水稻却能获得丰收。然而，在20余万亩的梯田中，没有一处水库和存水水塘，全靠天然灌溉系统满足梯田对水的要求。

山顶保留的植被，是紫鹊界梯田的主要水源地。紫鹊界梯田的表层土壤多为沙壤土，透水性能极好。山顶郁郁葱葱的各类树木，在雨季时，将自然降水截留并保存下来。水分透过土壤，缓缓向山腰中的梯田渗透。紫鹊界特殊的地质条件，使得山体的基岩裂隙和每一块构筑梯田的石块土坎，都有汩汩水流不断渗出。狭长的梯田田块还具有输水的功能，大部分梯田通过"借田输水"就能满足日常灌溉的需求。而天然的山谷溪沟，则是紫鹊界梯田的排水干渠，辅以人工开挖的短渠，并在梯田的适合位置开排水口，涝水和尾水就能够顺利排泄出来。

在世代的发展中，梯田稻作已深入苗族、瑶族人的心中，与他们传统的渔猎文化相融合，成就了紫鹊界人与自然和谐共处的稻作文化。

地理位置：北纬27°31′—28°14′、东经110°45′—111°41′，地处湖南省娄底市新化县。
气候特点：亚热带季风气候。
认定时间：2013年被农业部列为首批中国重要农业文化遗产，作为"中国南方山地稻作梯田系统"之一，于2018年被联合国粮农组织列为全球重要农业文化遗产。

Geographical location: 27°31′-28°14′N, 110°45′-111°41′E; located in Xinhua County, Loudi City, Hunan Province.
Climate type: subtropical monsoon climate.
Time of identification: In 2013, it was listed as one of the first batch of the China-NIAHS by the Ministry of Agriculture. In 2018, as one of the Rice Terraces in Southern Mountainous and Hilly Areas in China, it was listed as one of the GIAHS by the FAO.

The Fusion System of Southern Rice Culture and Miao-Yao Montanic Fishing & Hunting Culture
Xinhua Ziquejie Terraces

Among the terraced fields in the mountainous and hilly areas of southern China, the Ziquejie Terraces stand out from the competition with their unique irrigation system. The Ziquejie terraces are located in Xinhua County, Loudi City, west of central Hunan Province. Hills and slopes of Fengjia Mountain are covered with the terraces.

Since the Qin and Han Dynasties, farmers living in the Fengjia Mountain area have begun to build terraces. At that time, the Miao-Yao people escaped into the deep mountains of Fengjia Mountain in order to avoid the bullying of other powerful ethnic groups. In order to make a living, they opened up the wasteland and built terraces. In the Song and Ming Dynasties, the residents of Miao, Yao, Dong and Han had gathered in the area of Ziquejie and Fengjia Mountain. The population was so large, the land supply was in so short supply that the terraced area began to expand rapidly.

Even during the period of rapid expansion, people were not undisciplined on mountain-blasting and land reclamation. They preserved the vegetation on the top of the mountain and built terraces on the mountainside, while the flat parts at the foot of the mountain were reclaimed into contiguous farmland, forming an agroecological landscape of "a hat on the top, a belt around the hillside, a skirt at the foot of the mountain". This kind of layout had guaranteed the state that "the world is in chaos, there is no worry in this place; the world is full of drought, and there is harvest in this place".

The Ziquejie Terraces are located in the subtropical monsoon climate zone. Due to various reasons such as topography, the rainfall is unevenly distributed in time and space, and droughts have often visited. In a dry year, the rice fields at the foot of the mountain may encounter a total failure, but the rice in the terraced rice fields can be harvested. However, in more than 200,000 mu of terraced fields, there is no reservoir and storage ponds, relying on natural irrigation systems to meet the requirements of terraces for water.

The vegetation preserved on the top of the mountain is the main source of water for the Ziquejie terraces. The surface soil of the terraces of Ziquejie is mostly sandy loam, and the water permeability is excellent. The lush trees on the top of the mountain intercept and preserve rain water during the rainy season. Water penetrates through the soil and slowly into the terraces in the mountainside. The special geology of Ziquejie makes the bedrock fissures of the mountain body and every block of terraced field seep continuously. The narrow-terraced fields also bear the function of water delivery. Most of the terraces can meet the daily irrigation needs by "borrowing water". The natural valley creek is the main drainage channel of the Ziquejie terraces. Also, supplemented by artificially excavated short channels, and the drainage outlets opened at the appropriate positions of the terraces, the flood and tail water can be discharged smoothly.

In the development of generations, rice terraces have been deeply rooted in the hearts of Miao and Yao people, and they are integrated with their traditional fishing & hunting culture, which has made the rice culture of the people of Ziquejie and nature coexist harmoniously.

紫鹊界梯田是由森林、民居、梯田、水系交错组成的南方山地农业生态系统，属雪峰山余脉的奉家山地段。紫鹊界梯田依山就势而造，小如碟、大如盆、长如带、弯如月，形态各异、变化万千，宛如天上瑶池、人间仙境。图为朝霞映照下的紫鹊界梯田。（田辉举 摄）

Ziquejie Terraces is an agroecosystem in southern mountainous areas, which is composed of forests, residential dwellings, terraces and water systems. It belongs to Fengjia Mountain section of the stretching branches of Xuefeng Mountain. Terraces here are built based on the terrain. Some of them are as small as dishes, as big as salvers, as long as belts, and as curved as the moon. With various and changeable shapes, they are like fairylands in the paradise. The picture shows the Ziquejie terraces in the morning glow. Photographed by Tian Huiju

紫鹊界降水量大，森林覆盖率高，因此水系十分发达，水源充沛，村民只需在田边地角随意挖一缺口，就能见到并引流山泉水。这里凭借神奇独特的基岩裂隙孔隙水源，构成纯天然自流灌溉工程。潺潺流水，四季不绝，久旱不竭，洪涝无忧。山有多高，水有多高，田就有多高，这种自流灌溉系统堪称"世界水利灌溉工程之奇迹"。这里祖祖辈辈的生活用水也是基岩裂缝中渗出的山泉。（田辉举 摄）

Ziquejie has a large amount of rainfall and high forest coverage, which lead to developed water system and abundant water sources. Accordingly, the villagers only need to dig a gap in the corner of the field to see and drain the spring water. With the magical unique bedrock fissure pores, the water source makes a pure natural self-flow irrigation project. The murmuring flow of water is endless all over the four seasons, and even with the prolonged droughts there are no worries about floods. And how high the mountains are, how high the water is, so how high the fields are. This self-flowing irrigation system can be called "the miracle of the world's irrigation projects". The water used by our ancestors of generations for daily life is also the mountain spring seeping from the cracks in the bedrock. Photographed by Tian Huiju

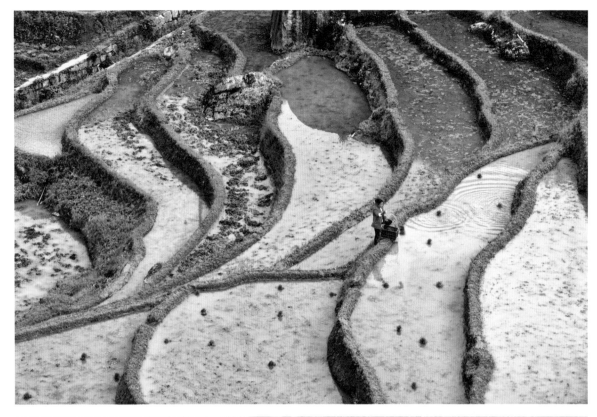

紫鹊界梯田依靠森林植被、土壤和田埂综合形成自然的储水保水系统，凭借神奇独特的基岩裂隙孔隙水源，构成纯天然自流灌溉工程。图为当地农民在田间抛秧苗，进行无公害种植。（田辉举 摄）

Ziquejie terraced fields rely on forest vegetation, soil and ridge to form a natural water storage and water conservation system, and rely on magical and unique bedrock fissure pore water source to form a pure natural self-flow irrigation project. The pictures show local farmers throwing seedlings and planting without pollution in the field. Photographed by Tian Huiju

华中地区 · Central China／湖南省 · Hunan Province

紫鹊界梯田始于秦汉，盛于宋明，至今已有 2000 余年历史，是苗、瑶、侗、汉等多民族历代先民共同创造的劳动成果，是南方稻作文化与苗瑶山地渔猎文化融合的历史遗存。由于梯田的地理特点，近年来除了机器耕田，很多农民至今还是使用传统的水牛耕田。（田辉举 摄）

Ziquejie terraced fields began in the Qin and Han Dynasties and flourished in the Song and Ming Dynasties. They have a history of more than 2,000 years. They are created by the ancestors of many ethnic groups such as Miao, Yao, Dong and Han. They are the historical relics of the integration of rice culture in South China and fishing & hunting culture in Miao-Yao Mountains. Due to the geographical characteristics of terraced fields, in recent years, in addition to machine farming, many farmers still use traditional buffalo farming. Photographed by Tian Huiju

南方稻作文化与渔猎文化的巧妙融合成就了紫鹊界人与自然和谐共处的稻作文化遗存，这里民风淳朴，苗瑶风俗世代相传。作为至今保存在湖南中部和西南部的古老原始土著文化——梅山文化的代表性地域，紫鹊界民居古色古香、颇有特色；草龙舞、傩面狮身舞等风俗表演更是原始神秘、别有风情。（田辉举 摄）

The ingenious combination of southern rice culture and fishing & hunting culture has made the remains of rice culture of local people in harmony with nature. The folk customs here are simple, and the Miao-Yao customs have been passed down from generation to generation. As the ancient primitive indigenous culture preserved in the central and southwestern parts of Hunan-representative areas of Meishan culture, Ziquejie residences are antique and quite distinctive; the folk performances such as grass dragon dance and the dance of lion with Nuo's face are even more original and mysterious. Photographed by Tian Huiju

紫鹊界的饮食文化丰富多样，农家自种自养的纯天然无污染的农家菜，包括各种时令蔬菜以及干菜，还有猪血丸子、冬笋、笋粑，以及农家自酿的米酒等。（田辉举 摄）

The food culture of Ziquejie is rich and varied. The local farmers grow natural and pollution-free vegetables, including various seasonal vegetables and dried ones. Besides, pig bloodballs, winter bamboo shoots, rice wine etc. are also made by farmers. Photographed by Tian Huiju

紫鹊界梯田是多民族祖先用毕生的精力和心血，经过世世代代的劳动创造出来的，现在各级政府正在采取多种形式大力保护和传承祖先留下的这一宝贵财富。（田辉举 摄）

Ziquejie terraced fields were created by multi-ethnic ancestors through their lifelong efforts and generations of labor. Now governments at all levels are taking various forms to actively protect and inherit the precious wealth left by their ancestors. Photographed by Tian Huiju

2014年9月16日，在韩国光州举行的第22届国际灌溉排水大会暨国际灌溉排水委员会（ICID）第65届国际执行理事会上，新化水利灌溉工程被列入首批世界灌溉工程遗产名录。同期列入世界灌溉工程遗产名录的还有四川乐山东风堰、福建莆田木兰陂及浙江丽水通济堰。图为远眺紫鹊界。（田辉举 摄）

On September 16, 2014, at the 22th International Irrigation and Drainage Conference and the 65th International Executive Council of the International Irrigation and Drainage Commission (ICID), held in Gwangju, South Korea, the Xinhua Water Irrigation Project was licensed to be included in the first batch of Heritage Irrigation Structures List. At the same time, Dongfengyan in Leshan, Sichuan, Mulanbeiin in Putian, Fujian and Tongji Weir in Lishui, Zhejiang were listed in the list. The picture shows Ziquejie in a distance. Photographed by Tian Huiju

世界原始稻作文化的"活化石"
湖南新晃侗藏红米种植系统

与常见的白色稻米不同，红米的外观正如其名——有着呈胭红色的外表。即使对其进行蒸制、发酵等加工后，红米也依然保持着这种颜色。红米是一种较为原始的水稻品种，因为水稻育种的不断演化以及特殊的历史原因，红米种植区越来越少。在湖南怀化新晃，种植红米的传统被当地的侗族人保留了下来。湖南新晃侗藏红米种植系统即为侗族、苗族民众以种植侗藏红米为特色的农业生态系统。

新晃位于湖南省的最西端，雪峰山以西，武陵山以南，地处我国亚热带地区，季风气候让这里四季分明、雨量充沛，境内土壤含有丰富的矿物质。当地曾有考古遗址出土过稻灰和稻谷颗粒遗迹，证实早在8000年前，就有人类在这片土地上从事稻作。

新晃是侗族聚居的地方，侗族占到了新晃县域总人口的80%。侗族是以农业耕种为主的民族，历史上一直生活在我国南方的水稻种植区。在种植水稻方面，侗族人有着悠久的历史和丰富的经验。

红米，作为水稻中珍贵而又难得的物种，一直在这里被侗家人保留了下来。编纂于清代的《晃州厅志》中有当地（晃州为古地名，曾辖现部分新晃区域）种植红米的相关记载：糯之种则有扫箕糯、光头糯、牛皮糯、红糯、白糯、种禾糯诸名。其中的"红糯"，指的就是红稻米。

在新晃，红米不仅是侗家人的口粮，还是他们的"精神食粮"。红米的种植在这里能够保留下来，在一定程度上也得益于此。侗家人把红米这种难得一见的稻米奉为"神米"。侗族是一个信仰多神的民族，红米在侗族的巫傩、祭祀等文化中多有得见，是这些文化活动中不可或缺的一种物料。

在山下的水田里种植水稻的同时，侗家人还在山坡上种植林木，形成了稻林相间的农业景观。此外，侗家人还遵循着插秧时节养鱼、收割以后养鸭的农业生产方式，形成了一种良性、丰富、循环的农业生态体系。

地理位置：北纬27°4'16"—27°29'58"、东经108°47'13"—109°26'45"，地处湖南省怀化市新晃侗族自治县。
气候特点：亚热带季风气候。
认定时间：2014年被农业部列为第二批中国重要农业文化遗产。

Geographical location: 27°4'16"-27°29'58"N, 108°47'13"-109°26'45"E; located in Xinhuang Dong Autonomous County, Huaihua City, Hunan Province.
Climate type: subtropical monsoon climate.
Date of Identification: In 2014, it was listed as one of the second batch of the China-NIAHS by the Ministry of Agriculture.

The "Living Fossil" of the World's Original Rice Culture

Xinhuang Red Rice Planting System of Dong Nationality

Unlike the common white rice, the red rice looks just like what its name shows– it has a crimson appearance. Even after steaming, fermenting, etc., the red rice still retains this color. Red rice is a relatively primitive rice variety. Due to the continuous evolution of rice breeding and special historical reasons, there are fewer and fewer red rice growing areas. In Xinhuang County, Huaihua City, Hunan Province, the tradition of planting red rice has been preserved by the local Dong people. The Red Rice Planting System of Dong Nationality is an agro-ecosystem characterized by the red rice cultivation by local Dong and Miao people.

Xinhuang is located at the western end of Hunan Province, west of Xuefeng Mountain and south of Wuling Mountain. It is located in the subtropical region of China. The monsoon climate makes the four seasons distinct and the rainfall is abundant. The soil in the territory is rich in minerals. The unearthed remains of rice ash and grain at the local archaeological sites prove that as early as 8,000 years ago, people were engaged in rice cultivation on this land.

Xinhuang is a place where Dong people live together. Dong people account for 80% of the total population of Xinhuang County. Dong nationality is an ethnic group mainly engaged in agricultural cultivation. It has been living in rice planting areas in the south of China in history. Dong people have a long history and rich experience in planting rice.

Red rice, a precious and rare species in rice, has been preserved here by the Dong people. *Huangzhou Hall Chronicle*, compiled in the Qing Dynasty, contains relevant records of planting red rice in local areas (Huangzhou is an ancient place name, which once governed part of today's Xinhuang area). The species of glutinous rice are sweeping dustpan glutinous rice, bald glutinous rice, cowhide glutinous rice, red glutinous rice, white glutinous rice and grain glutinous rice. Among them, red glutinous rice refers to red rice.

In Xinhuang, red rice is not only the ration of Dong people, but also their "spiritual food". Red rice planting can be preserved here, to some extent, also benefited from this. The Dong people regard red rice, a rare kind of rice, as "divine rice". As a multi-god people, red rice can be much seen in Dong culture such as sorcery nuo, sacrifice and so on. It is an indispensable material in these cultural activities.

When rice was planted in paddy fields at the foot of the mountains, the Dong people planted trees on the slopes, forming an agricultural landscape with rice forests. In addition, the Dong people also follow the agricultural production mode of raising fish during transplanting season and ducks after harvesting, forming a benign, rich and recycling agricultural ecological system.

新晃侗藏红米种植系统有悠久的历史文化底蕴。新晃气候温和，光照充足，严寒期短，无霜期长，境内土壤里含有大量的铁、锌、钙等多种微量元素，为侗藏红米提供了独特的生长环境。图为扶罗基地核心区平坝稻田，稻田与侗寨、蓝天完美统一。（新晃县委宣传部 供图）

The Xinhuang Red Rice Planting System of Dong Nationality has a long history with cultural connotations. Xinhuang has a mild climate, sufficient sunshine, short cold period and long frost-free period. The soil contains a large amount of trace elements such as iron, zinc and calcium, which provides a unique environment for the growth of red rice. The picture shows the Pingba paddy field in the core area of the Fuluo base. The rice fields are perfectly integrated with the Dong village and the blue sky. Provided by Publicity Department of Xinhuang County Party Committee

湖南新晃侗藏红米种植系统是新晃侗乡数千年来农耕文明的历史传承。红米稻种更是珍贵的、难得的物种资源。千百年来，侗藏红米凭着独特的人文地理环境和栽培习俗，在杂交水稻发祥地湖南怀化的新晃侗乡得以保存下来。上图为红米稻田和侗寨交错其间，和谐共生；下图则是侗藏红米种植基地稻田里的"侗藏红米"艺术字，彰显了当地稻农对于红米的热爱。（新晃县委宣传部 供图）

The Xinhuang Red Rice Planting System of Dong Nationality is the historical inheritance of farming civilization for thousands of years. Red rice is a precious and rare species. For thousands of years, it has been preserved where hybrid rice originated, because of its unique human and geographical environment and cultivation customs. The picture above shows the harmonious symbiosis between red rice paddy field and Dong village. The picture below shows the art words "red rice of Dong" in the field of the red rice planting base, which highlights the local rice farmers' love for red rice. Provided by Publicity Department of Xinhuang County Party Committee

新晃侗藏红米种植系统体现了生态农业和循环农业的理念。山上封山育林，山下引水灌溉，林稻相间，相辅相成。水旱轮作的循环系统，既丰富了农作物的种植结构，又改善了土壤的营养成分。水稻种植与养鱼、养鸭的有机结合，无形中建立了一套良性循环的农业生态体系。图为新晃侗乡高山原始林木和红米稻田的稻田养鱼。（新晃县委宣传部 供图）

The Xinhuang Red Rice Planting System of Dong Nationality embodies the concepts of ecological circular agriculture. The hills are closed for planting trees, the water is diverted to irrigated under the mountain, and the forests and rice are mutually complementary. The circulatory system of water and drought rotation not only enriches the planting structure of crops, but also improves the nutrient composition of the soil. The organic combination of the rice paddy planting system and the fish and duck breeding has invisibly established a benign cycle of agroecological system. The pictures show the alpine virgin forests and fishing farming in the red rice fields in Xinhuang Dong Autonomous County. Provided by Publicity Department of Xinhuang County Party Committee

侗藏红米除含有丰富的微量元素以及植物性蛋白质、植物性脂肪外，还富含 B1、B2、B6 等多种维生素和 18 种人体必需的氨基酸。图为成熟的侗藏红米和红米粥、红米饭、红米酒、红米粑、红米粽子等红米制食品。（新晃县委宣传部 供图）

In addition to rich trace elements, plant protein and vegetable fat, red rice of Dong is also rich in vitamins such as B1, B2, B6 and 18 essential amino acids. The pictures show mature red rice and red rice porridge, red rice, red rice wine, red rice cakes, red rice dumplings and other red rice food. Provided by Publicity Department of Xinhuang County Party Committee

侗藏红米不仅是侗家人的食粮,更是侗家人崇尚自然的精神支柱,被侗家人视为"神米",与巫傩文化、祭祀文化等侗民俗文化有着密切的联系。傩,古书解为驱鬼逐疫。古老的图腾崇拜和鬼神信仰,使我们的祖先总是习惯借助于"傩"这种神秘力量来实现自己的美好愿望,并辅之以歌舞,这便是最初的傩戏。新晃的傩戏"咚咚推",多年前就被国内外专家学者称为"中国戏剧的活化石",它是巫傩文化的早期产物。(新晃县委宣传部 供图)

The red rice is not only the food of the Dong People, but also the spiritual pillar of the Dong people's worship of nature. It is regarded as the divine rice by them, and closely connected with the culture of sorcery nuo and sacrificial rites. Nuo is interpreted as driving away ghosts and epidemics in the ancient books. The ancient totem worship and belief in ghosts and gods made our ancestors always get used to using the mysterious power of Nuo to achieve their good wishes, supplemented by singing and dancing, which was the origin of Nuo opera. Xinhuang's Nuo opera-dongdongtui, which was called "the living fossil of Chinese drama" by experts and scholars at home and abroad many years ago, is an early product of sorcery nuo culture. Provided by Publicity Department of Xinhuang County Party Committee

新晃侗藏红米种植系统有 8000 年的历史传承。在现代农业技术影响力越来越大的今天，保留丰富的稻种资源不仅对丰富我国稻类遗传资源、稻作生产、品种改良、稻作科学研究及生态安全维护有着积极的作用，也是解决人类未来粮食安全的物质保证。图为机械化收割红米和太阳能灭虫灯守护的侗藏红米稻田。（新晃县委宣传部 供图）

The Xinhuang Red Rice Planting System of Dong Nationality has a history of 8,000 years. Today, with the increasing influence of modern agricultural technology, the preservation of abundant rice resources is not only positive for enriching China's rice genetic resources, rice production, variety improvement, rice research and ecological security but also a material guarantee for solving the future food security of mankind. The pictures show the red rice paddy field protected by mechanized harvesting of red rice and solar insect control lamps. Provided by Publicity Department of Xinhuang County Party Committee

华南地区·South China

广东省

岭南茶文化代表——广东潮安凤凰单枞茶文化系统

广西壮族自治区

壮瑶人民农耕文明的结晶——广西龙胜龙脊梯田农业系统

岭南茶文化代表
广东潮安凤凰单枞茶文化系统

岭南地区气候炎热，人们日常活动中的需水量极大。因茶叶可以生津解渴，故饮茶对于岭南地区的人们来说，是生活中一项必不可少的活动。岭南人喝茶是出了名的精细，这种精细不仅体现在茶具和饮茶方式中，更体现在茶的种植上——潮安凤凰单枞茶便是典型例证。与其他茶叶批量采摘和炮制不同的是，凤凰单枞茶实行的是单株采摘、单株制茶以及单株销售的生产体系。广东潮安凤凰单枞茶文化系统即以凤凰单枞茶种植和加工为特色的茶文化系统。

凤凰单枞茶的主产地凤凰镇位于潮州市潮安区的山区中。虽然处于亚热带气候区，但是这片区域群山环绕、云遮雾罩，为茶树的生长提供了凉爽、湿润的生长环境。凤凰单枞茶生长在海拔600—1200米的山坡上，山坡上的表层土壤以呈酸性的红壤为主，单枞茶在这里找到了自己的天地。在凤凰单枞茶的产区，有超过3700株树龄在200年以上的古茶树，被认为是"世界罕见的优稀茶树资源"。

凤凰茶叶种植历史悠久，传说始于南宋末年，文献记载见诸嘉靖二十六年（1547年）的《潮州府志》，而凤凰产茶的历史远早于此。南宋时期，生活在凤凰附近的农民发现用茶叶泡水有生津止渴等多种效用，于是开始对当地的茶树进行选种和广泛栽培。明代时，凤凰产的茶叶凭借自身上乘的品质，被选为皇家的贡品。凤凰茶得名"单枞茶"则是在清代。茶农发现实行单株生产，可以对茶树中的优秀品种进行选择，于是便开始使用此种生产方式，产出的茶叶被称为"凤凰单枞茶"。

凤凰单枞茶属于乌龙茶类。鲜叶采摘后，需要经过晒青—晾青—做青—杀青—揉捻—烘焙六道工序，才能制成成茶。独特的地理环境和生产模式，使凤凰单枞茶有着"一树一香"的特性。茶叶是凤凰的主要产业，制成凤凰单枞茶的六道工序在凤凰传承了数十代人，也养活了凤凰的数十代人。凤凰单枞茶曾通过海上丝绸之路远销海外，凤凰也因茶而闻名，成为名茶之乡。

地理位置：北纬23°43′40″—23°41′17″、东经116°39′47″—116°14′09″，地处广东省潮州市潮安区。
气候特点：亚热带海洋性季风气候。
认定时间：2014年被农业部列为第二批中国重要农业文化遗产。

Geographical location: 23°43′40″-23°41′17″N, 116°39′47″-116°14′09″E; located in Chao'an District, Chaozhou City, Guangdong Province.
Climate type: subtropical marine monsoon climate.
Time of identification: In 2014, it was listed as one of the second batch of the China-NIAHS by the Ministry of Agriculture.

A Representative of Lingnan Tea Culture
Chao'an Phoenix Single Cluster Tea Culture System

The climate is so hot in Lingnan area that people need much water in their daily activities. Since tea has the potent effect in quenching thirst, drinking tea is a crucial part of daily life for local people. They are famously refined when drinking tea, which is not only reflected in selecting tea set and drinking tea, but also in tea planting. Phoenix Single Cluster Tea at Chao'an County in Guangdong Province is a representation. Instead of batch picking and processing, Phoenix Single Cluster Tea is manufactured within a system in which single tea plant is picked, processed, and sold. Chao'an Phoenix Single Cluster Tea Culture System is one characterized by the cultivation and processing of Phoenix Single Cluster Tea.

Phoenix Town, the main producing area of Phoenix Single Cluster Tea, is located in the mountainous area of Chao'an District, Chaozhou City. Although it is in a subtropical climate zone, this area is surrounded by mountains and clouds, providing a cool and humid environment for the growth of tea trees. Phoenix Single Cluster Tea grows on the slopes of 600-1200 meters above sea level. The dominant surface soil on the slopes is acidic red soil, for which Single-cluster tea finds its own world here. In the whole production area, there are more than 3,700 ancient tea trees with estimated ages of more than 200 years, which is considered to be "a precious and rare tea tree resource in the world".

Phoenix tea planting has a long history. According to legend, it began from the late Southern Song Dynasty. Documents related were recorded in *Local Chronicles of Chaozhou* records of the 26th year of the Jiajing Period (1547), but actually tea production in Phoenix began much earlier than that. During the Southern Song Dynasty, farmers living nearby found that soaking water with tea had many functions, such as promoting the production of body fluids and stopping thirst. Then they began to select and cultivate local tea trees extensively. In the Ming Dynasty, the tea produced here was even selected as a royal tribute by virtue of its superior quality. It was until the Qing Dynasty that Phoenix tea was named "single- cluster tea". At that time, tea growers found it favorable to select excellent varieties of tea through single-plant production. Ever since this method has been adopted, and the tea produced is called "Phoenix Single Cluster ".

Phoenix Single Cluster Tea belongs to oolong tea. After fresh leaves are picked, it is necessary to go through the process of sunshine drying, air drying, making green, water removing, rolling and baking to make tea. The unique geographical environment and production mode enables the Phoenix Single Cluster Tea to have the feature of "one tree and one fragrance". Tea is the main industry of Phoenix Township, whose six procedures of making Phoenix Single Cluster Tea have been passed down through tens of generations, and have also supported tens of generations. Phoenix Single Cluster Tea has been sold overseas via the Maritime Silk Road, for which Phoenix has become a famous tea-producing town.

广东潮安凤凰单枞茶文化系统位于潮州市北部山区的凤凰镇,这里群山环抱,云蒸雾蔚,山川秀美,最高海拔达 1498 米。由于这里的气候和土壤适合于茶树喜温、喜酸、好湿和耐阴的生物学特性,于是便构成了生产名优茶得天独厚的自然环境条件。图为今天的凤凰镇茶山。(张伟雄、谢德楠 摄)

Phoenix Single Cluster Tea Culture System in Chao'an City, Guangdong Province is located in Phoenix Township in the northern mountain area of Chaozhou City, where mountains are surrounded by clouds and mists, beautiful mountains and rivers, and with a maximum altitude of 1498 meters. With climate and soil suitable for biological characteristics of tea plants, such as tolerance of heat, acidity, humidity and shade, it makes the unique natural environment for producing famous and excellent tea. The picture shows today's tea mountains in Phoenix Township. Photographed by Zhang Weixiong and Xie Denan

华南地区・South China／广东省・Guangdong Province

据称,凤凰镇茶山上有一棵千年茶树王,生长在海拔高度约1300米的乌岽山李仔坪村,系南宋末年村民李氏几经选育繁殖的一株茶树,是当今凤凰名茶中最古老的一棵,更因树老、种奇、香异而驰名古今中外。(张伟雄 摄)

Allegedly there is one Tea King with a history of nearly one thousand years, growing in Lizaiping Village on Wudong mountain about 1300 meters above sea level. It is said that it is a tea tree selected and bred by Lishi, a villager in the late Southern Song Dynasty, and it is the oldest one among the famous phoenix tea today. Because of its oldness, peculiar and unique fragrance, it is well-known in China and the world. Photographed by Zhang Weixiong

华南地区 · South China / 广东省 · Guangdong Province

单枞茶系在凤凰水仙群体品种中选拔优良单株茶树，经培育、采摘、加工而成。单枞茶实行分株单采，当新茶芽萌发至小开面时（即出现驻芽），即按一芽、二三叶标准，用骑马采茶手法采下，轻放于茶萝内。有强烈日光时不采，雨天不采，雾水茶不采的规定。图为当地茶农采茶中。（张伟雄、谢德楠 摄）

Single-cluster tea is made in the following steps: First excellent single tea plant is selected from the genus of phoenix narcissus, and then cultivated, harvested and processed. Single-cluster tea leaves are picked on a basis of individual ramet. When the new tea buds are germinated to a small open surface (a bud), with the standard of one bud and two three leaves, they are harvested carefully placed in the tea sieves. There is a rule that there is no strong sunlight, no rain and no fog when picking tea leaves. The picture shows the local tea farmers picking tea. Photographed by Zhang Weixiong and Xie Denan

凤凰单枞茶的采摘初制工艺,是手工或手工与机械生产相结合。其制作过程有晒青—晾青—做青—杀青—揉捻—烘焙六道工序,环环相扣,每一工序都不能粗心随意。(张伟雄、谢德楠 摄)

The initial picking process of Phoenix single cluster tea is a combination of manual or manual and mechanical production. Its production process includes six procedures: sunshine withering, air drying, making green, water removing, rolling and baking. The six steps are closely related, and each should be performed with much care. Photographed by Zhang Weixiong and Xie Denan

潮州人讲究"茶道","功夫茶"是"潮州人习尚风雅,举措高超"的象征。冲泡功夫茶需要特制的泥炉和橄榄木炭。橄榄木炭即其中黑色橄榄状物品,为煮冲单枞茶水之用。(谢琳 摄)

Chaozhou people pay attention to the "tea ceremony", and "Kung Fu Tea" is a very symbol of "Chaozhou people are elegant in customs and superb in actions" Brewing Kung Fu Tea requires a special clay oven and olive charcoal. Olive charcoal is the black olive-like object, which is used to cook single-cluster tea. Photographed by Xie Lin

因成茶香气、滋味的差异，当地习惯将单枞茶按香型分为黄枝香、芝兰香、桃仁香、玉兰香等多种，据说潮州单枞茶每一道茶的最后几滴是精华。（谢琳 摄）

Due to the difference in tea aroma and taste, local single-cluster tea varieties are customarily divided into Huangzhixiang, Zhilanxiang, Taorenxiang and Yulanxiang, etc. based on the fragrance types. It is said that the last few drops of Chaozhou single-cluster tea are its essence. Photographed by Xie Lin

壮瑶人民农耕文明的结晶
广西龙胜龙脊梯田农业系统

广西龙胜各族自治县位于广西壮族自治区东北部、湘桂交界之处。如果不是龙脊梯田，这个深藏于桂北山区的小城，恐怕还鲜为人知。

地处越城岭山脉西南麓的龙胜县有着"万山环峙、五水分流"的地理特征。这里山高水急，沟幽谷深。全县有21座海拔在1500米以上的山峰，龙脊梯田便隐藏在这深山之中。

恶劣的自然地理环境、少有适合农业耕作的土地，使得这里在元代以前偏僻荒凉、人烟稀少。元代时，壮族和瑶族先民陆续迁居到此，他们把稻作文明也带到了这里。居住在这里的少数民族先民用"刀耕火种"开山造地，把坡地整为梯地，待田块逐渐定型后，再灌水犁田种植水稻，形成从山脚盘绕到山顶"小山如螺，大山成塔"的壮丽梯田景观。

在梯田开创的初期，壮、瑶族人把梯田建在山腰以下，壮寨和瑶寨村庄建在山腰上，山顶的森林则被保留了下来。

这个原则一直被遵循着。人们深知森林对于梯田来说有多么重要。湿润的亚热带季风气候保证了龙胜山区有充沛的雨量供给，这些被保留下来的植被便是天然降水的存水储水装置和调节装置。茂密的植被将天然降水保留下来，并缓冲了山顶水流的速度，增强了雨水向地下渗透的能力。汩汩清泉从林中潺潺而下，在龙脊梯田区形成"山有多高，水有多高"的景观。

当地农民通过封山育林等手段，对山顶的原始森林严加保护，同时，梯田周边的植被也获得了良好的保护，随山势海拔变化分布着不同的乔、灌、草等植被类型。

现在，除了水稻外，当地壮、瑶居民还根据海拔差异种植辣椒、红薯、芋头等普通作物和茶叶、罗汉果等作物，并在田间养殖凤鸡、翠鸭等家禽，保存和培育了龙胜地区丰富的农业种质资源。

地理位置：北纬25°35′—26°17′、东经109°32′—100°14′，地处广西壮族自治区桂林市龙胜各族自治县。

气候特点：亚热带季风性湿润气候。

认定时间：2014年被农业部列为第二批中国重要农业文化遗产，作为"中国南方山地稻作梯田系统"之一于2018年被联合国粮农组织列为全球重要农业文化遗产。

Geographical location: 25°35'-26°17'N, 109°32'-100°14'E; located in Longsheng County, Guilin City, Guangxi Zhuang Autonomous Region.

Climate type: subtropical monsoon humid climate.

Date of identification: In 2014, it was listed as one of the second batch of the China-NIAHS by the Ministry of Agriculture. In 2018, as one of the Rice Terraces in Southern Mountainous and Hilly Areas in China, it was listed as one of the GIAHS by the FAO.

The Crystallization of Farming Civilization of Zhuang and Yao Nationalities
Longsheng Longji Terraces

Longsheng County is located in the northeast of Guangxi Zhuang Autonomous Region and at the junction of Hunan and Guangxi. If it weren't for the Longji Terraces, this small county hidden deep in the mountainous areas of northern Guilin would probably be unknown.

Located in the southwestern foot of the Yuechengling Mountains, Longsheng County features "being surrounded by ten thousand mountains with five rivers flowing through". And those mountains are high, torrents are rapid and the valleys are deep. There are altogether 21 peaks over 1,500 meters above sea level in the county, and the Longji Terraces are hidden in the deep mountains.

The harsh natural conditions of geographical environment and the lack of land suitable for agricultural farming made it remote and deserted and sparsely populated before the Yuan Dynasty. Later in the Yuan Dynasty, the Zhuang and Yao nationalities moved to this place, where they brought the rice civilization. The ethnic minority ancestors living here reclaimed the mountains as farmland by the slash-and-burn practices, and transformed the slopes into terraces. After the fields were gradually shaped, the rice fields were planted and ploughed, forming the amazing terraced landscape of " hills like conches and mountains like towers" from the foot to the top of the mountain.

In the early stage of terrace creation, the Zhuang and Yao people built terraces below the mountainside, the villages on the mountainside, and the forests at the top of the mountain was naturally preserved.

This principle has been followed by local people, for they know how important forests are to terraces. The humid subtropical monsoon climate guarantees ample rainfall supply in the Longsheng Mountain area. This preserved vegetation is a natural device for storing and regulating water. The dense vegetation acts as buffer strips to help preserve the rainfall runoff and control the mountain torrents, enhancing the ability of the rain to penetrate into underground. The gurgling spring water flows through the forest, forming a landscape of "how high the mountains are and how high the water is (the water is as high as the mountains are)" in the Longji terraced area.

Local farmers have severely protected the virgin forests on the top of the mountain by means of closing hillsides for afforestation. At the same time, the surrounding vegetation of the terraced fields has also been well conserved. The stereoscopic climate formed with the change of mountain elevation distributes different vegetation types like trees, shrubs and grasses.

Now, in addition to rice, the local residents of Zhuang and Yao ethnic groups grow common crops such as pepper, sweet potato, taro, tea and momordica grosvenori(luohanguo) based on altitude differences. And they also raise poultry such as phoenix chickens and green ducks in the fields. All their efforts have effectively conserved and cultivate rich agricultural germplasm resources in Longsheng area.

龙胜龙脊梯田系统是以壮、瑶文化和壮美景观为特色的南方山地梯田生态系统。悠久的历史，良好的生态，丰富的种质资源，蔚为壮观的梯田景观和独特的壮、瑶民俗风情使龙脊梯田的自然生态与民族文化得到了高度融合，声名享誉中外。（杨广森 摄）

The Longsheng Longji Terraces is a southern mountain terrace ecosystem characterized by Zhuang-Yao culture and magnificent scenery. With a long history, good environment, rich germplasm resources, spectacular terraced landscape and unique Zhuang and Yao folk customs, the natural ecology and national culture of Longji Terraces have been highly integrated, and its reputation is well-known both at home and abroad. Photographed by Yang Guangsen

龙脊梯田地处亚热带，四季分明。梯田所在山脉山高谷深，落差巨大，海拔最高处1850米，最低处只有300米。山顶是大面积的原始森林和次生林，森林下方是规模宏大的梯田，壮寨和瑶寨散布在山腰上。图为春、夏、秋、冬四季的龙脊梯田。（韦毅刚 摄）

Longji Terraces is located in the subtropical zone with four distinct seasons. The terraced mountains are high and the valleys are deep. The highest peak is as high as 1,850 meters and the lowest is only 300 meters. The top of the mountain is a large area of virgin forest and secondary forest. Under the forest is a large terraced field, and Zhuang and Yao villages are scattered on the mountainside. The pictures show the Longji Terraces in the spring, summer, autumn and winter seasons. Photographed by Wei Yigang

800多年来，龙脊梯田已融入了当地居民的生活与文化的各个方面。这里保存着以梯田农耕为代表的稻作文化、以干栏民居为代表的建筑文化、以铜鼓舞和弯歌为代表的歌舞文化和以"龙脊四宝"为代表的饮食文化，构成了龙脊梯田独具特色的文化吸引力。图为龙脊春耕和龙脊梳秧节期间的大型梯田火把夜景。（文雪山 摄）

For more than 800 years, Longji terraces have been integrated into all aspects of the lives and culture of local residents. Here, there is a unique rice culture represented by terraced rice farming, an architectural culture represented by bamboo houses, a song and dance culture featuring bronze-drum dance and a bending song, and a food culture represented by "four treasures of Longji". The pictures show the night view of the large terraced torches during the Longji Spring Plowing and Longji Combing Seedling Festival. Photographed by Wen Xueshan

独特的地理和生态条件使得龙脊梯田产出丰富。当地壮、瑶居民根据海拔差异因地制宜地种植水稻、辣椒、红薯、芋头等普通作物和茶叶、罗汉果，养殖凤鸡、翠鸭等地理标志性农副产品，保存和培育了丰富的农业种质资源。图为瑶寨晾晒作物的场景。（马红专 摄）

The unique geographical and ecological conditions make it possible for the Longji terraces to be rich in farm produce. Based on altitude differences, local Zhuang and Yao residents have planted common crops like rice, pepper, sweet potato, taro etc. and geographic agricultural and sideline products such as tea, momordica grosvenori(luohanguo), phoenix chicken and green duck. As a result, rich agricultural germplasm resources have been effectively preserved and cultivated. The picture shows the scene of Yao people harvesting and drying crops. Photographed by Ma Hongzhuan

农历六月六是民间传统的节日，各地叫法不同，有叫"半年节"的，有叫"尝新节""晒衣节"的，"晒衣节"得名于每逢农历六月六这天村民都要将自己家的花衣、花裙和装饰用具等出来晾晒，每家每户的晒排和晾衣架上红红绿绿的衣服，形成了极具特色的民间传统节日景象。在龙脊，除春节外最隆重的一个传统节日就是晒衣节。（马红专 摄）

June 6 of the lunar calendar is a traditional folk festival. There are different names from place to place. There are Six Months Festival and Tasting New Festival. The Sun-drying Festival is also one of them. On June 6, villagers need to take out their own clothes, skirts and decorative utensils to sun them. The sun-drying arrangement of each household and the red-green clothes on the racks form a distinctive folk festival, and was named accordingly. For people in Longji, it is the second most important traditional festival behind the Spring Festival. Photographed by Ma Hongzhuan

西南地区・Southwest China

四川省

川西北山地药用植物文化的代表——四川江油辛夷花传统栽培体系

云南省

山区稻作梯田的典型代表——云南红河哈尼稻作梯田系统

世界茶树原产地和茶马古道起点——云南普洱古茶园与茶文化系统

传统核桃与农作物套作农耕模式——云南漾濞核桃—作物复合系统

云南壮族稻作文化的代表——云南广南八宝稻作生态系统

3000年水旱轮作的"活化石"——云南剑川稻麦复种系统

贵州省

传统稻鱼鸭共生农业生产模式——贵州从江侗乡稻鱼鸭系统

川西北山地药用植物文化的代表
四川江油辛夷花传统栽培体系

辛夷花又被称作"木兰"或"紫玉兰"。人们常常分不清与它同属木兰科的玉兰花。与玉兰花相比,辛夷花的花色更加鲜艳、花朵更加小巧。辛夷花在中国有2000多年的栽种历史。诗人屈原的诗歌中,多处可见辛夷花:"朝饮木兰之坠露兮,夕餐秋菊之落英"(《离骚》);"桂栋兮兰橑,辛夷楣兮药房"(《楚辞·九歌·湘夫人》)……这说明,在屈原生活的战国时期,辛夷花就已经是一种常见的花木。

辛夷是一味中药材,它的树皮、叶、花蕾均可以入药。其中,辛夷花蕾对治疗鼻科方面的疾病有着明显的疗效。《本草纲目》中有这一方面的记载:"……辛夷之辛温走气而入肺,能助胃中清阳上行通于天,所以能温中治头面目鼻之病。"

辛夷花最初分布在中国云南、福建、湖北、四川等地,生长在海拔300—1600米的山坡上。经过长时间的人工选育之后,现在在我国各地均有栽种。但是,没有一个地方,像四川江油的吴家后山一样形成辛夷花海之势。

在江油市大康镇旱丰村的吴家后山,保有6万多棵古辛夷树,树龄最长的达400年。每年三四月份的辛夷花期,吴家后山都会形成颜色各异的辛夷花海60多处。四川江油辛夷花传统农业系统即以药用辛夷花树种植利用为特色的复合农林生态系统。

吴家后山对辛夷花的种植,可以追溯到清代康熙年间(1662—1722年)。当时,吴三桂家族为避乱来到这片区域隐居,并开始在这里种植辛夷花。后来,这片区域吴姓的人越来越多,便取名吴家后山。辛夷花也成了吴姓家族的一种标志,家族中形成了祭祀、节庆参拜辛夷树王和栽种辛夷祈福的传统。在后来的时间里,人们以辛夷花为中心,发展出了栽种辛夷树,采摘辛夷花,辛夷花树下种植天麻、百合、乌药,林间养蜂,放养山鸡和牛羊的农业生产方式。

现在的吴家后山,植被茂盛、种类繁多,生态环境良好,动植物资源极为丰富,被当地人称为"江油神农架"。

地理位置:北纬31°44′—31°52′、东经104°39′—104°44′,地处四川省绵阳市江油市。
气候特点:亚热带季风性湿润气候。
认定时间:2014年被农业部列为第二批中国重要农业文化遗产。

Geographical Location: 31°44-31°52′N, 104°39′-104°44′E; located in Jiangyou City, Mianyang City, Sichuan Province.
Climate type: subtropical monsoon humid climate.
Time of identification: In 2014, it was listed as one of the second batch of the China-NIAHS by the Ministry of Agriculture.

A Representative of Typical Alpine Traditional Chinese Medicine in Northwest Sichuan
Jiangyou Traditional Magnolia Cultivation System

Xinyi flower is also known as Flos Magnoliae Liliflorae or purple magnolia. People often confuse it with Yulan magnolia flower that belongs to the family Magnoliaceae. Compared with the latter, Flowers of Flos Magnoliae Liliflorae are brighter and smaller. They have been planted in China for more than 2000 years. In the poems by Qu Yuan, a poet and minister who lived during the Warring States period of ancient China, there are many places where the flowers of Magnolia can be seen. For example, in *Lisao*, he wrote "drinking the dew of Mulan in the morning, eating the fallen autumn chrysanthemum in the evening", and in the *Chuci (Elegies of Chu· Nine songs · Mrs. Xiang")*, he said, "Osmanthus wood is the crossbeam and orchid is the rafter, Magnolia wood is the sub-beam, and angelica dahurica decorates the side room." ...This shows that Magnolia was already commonly seen in the Warring States period when Qu Yuan lived.

Magnolia is a traditional Chinese medicine. Its bark, leaves and buds can be used as medicine. Among them, Flower Bud of Flos Magnoliae Liliflorae has obvious curative effect on the treatment of nasal diseases. In the *Compendium of Materia Medica*, there are records about its effect: "... (once taken) Xinyi's pungent flavor and warmth will enter the lungs and help send sufficient clear Qi from the stomach up, treating diseases of the heads, faces, noses or the eyes."

Flowers of Magnolia originally distributed in Yunnan, Fujian, Hubei, Sichuan and other places in China, growing on the slopes of 300-1600 meters above sea level. After a long period of artificial breeding, it is now planted in many places in China. However, there is no place like Wujiahoushan in Jiangyou, Sichuan Province, where the sea of Magnolia flowers is formed.

In Wujiahoushan where Hanfeng Village in Dakang Town of Jiangyou City is located, there are more than 60,000 ancient Magnolia trees, the longest of which is 400 years old. Every year in the March and April of the Magnolia flower period, Wujiahoushan will form more than 60 seas of Magnolia flowers of different colors. The traditional cultivation system of Magnolia in Jiangyou County of Sichuan Province is a complex agroforestry ecosystem characterized by the cultivation and utilization of Chinese medicine Magnolia flower and Magnolia trees.

The cultivation of Magnolia by villagers in Wujiahoushan can be traced back to the Kangxi period of the Qing Dynasty. At that time, the Wu Sangui family came to this area to escape from the chaos and began planting magnolia flowers here. Later, there were more and more people surnamed Wu here, for which this area was named Wujiahoushan. Magnolia had also become a symbol of the Wu family. It has formed a tradition of worshipping ancestors, paying homage to the king of magnolia tree and blessing by planting magnolia trees. In the later period, people developed the agricultural production mode of planting magnolia trees, picking magnolia flowers, planting gastrodia elata, lily, black peony, forest beekeeping, as well as stocking pheasants, cattle and sheep.

Now the Wujiahoushan has lush vegetation, a wide variety of species, a good ecological environment, and extremely rich animal and plant resources. It is called "Jiangyou Shennongjia" by the locals.

四川江油辛夷花传统农业系统位于江油市大康镇旱丰村吴家后山，面积约25平方千米，辛夷花种植面积近2万亩。系统内植被和生态环境良好，动植物资源和旅游资源极为丰富。图为吴家后山的辛夷花林。（凌弘 摄）

The Jiangyou Traditional Magnolia Cultivation System in Wujiahoushan, Hanfeng Village, Dakang Town, Jiangyou City, covers an area of about 25 square kilometers and totals nearly 20,000 mu. The vegetation and ecological environment in the system are good, and animal & plant resources and tourism resources are extremely rich. The picture shows the Magnolia flower forest in Wujiahoushan. Photographed by Ling Hong

吴家后山现存古辛夷树 6 万余株,垂直分布在吴家后山腹地,其栽培历史久远、花色品种齐全、规模宏大,已成为全国最大的辛夷花基地。图为百年辛夷树。(凌弘 摄)

There are more than 60,000 ancient Magnolia trees in Wujiahoushan, which are vertically distributed in the hinterland of the area. With a long history of cultivation, and diversified flowers and rich colors, it has become the largest Magnolia flower base in the country. The picture shows some magnolia trees of a hundred years old. Photographed by Ling Hong

它与玉兰同科，花色却远比玉兰鲜艳；艳若桃花，桃花却不如她庄重、恬静，它就是绽放于高山的辛夷花。（徐卫 摄）

Falling into the same family as Yulan Magnolia, but her flowers are much brighter than Yulan Magnolia. She is as beautiful as any peach blossom, but more solemn and quiet than them. She is nothing but the Magnolia blooming in the mountains. Photographed by Xu Wei

自古以来，吴家后山的村民们栽种辛夷树、采摘辛夷花，林下种植天麻、百合、乌药，林间养蜂、放养山鸡和牛羊等传统耕作方式一直延续至今。图为村民上树采摘辛夷花和销售辛夷花及树下产品。（凌弘 摄）

Since ancient times, the villagers of Wujiahoushan have adopted the traditional farming practices like planting magnolia trees, picking magnolia flowers, planting Gastrodia elata, Lily, black peony, forest beekeeping, and stocking pheasants, cattle and sheep. The pictures show the villagers picking and selling the flowers of Mangolia as well as the above-mentioned products under the Mangolia trees. Photographed by Ling Hong

吴家后山独特的地理位置和自然条件，造就了辛夷花独特的品质。辛夷花、树皮入罐为药，上桌为膳，有养生治病之功效。图为辛夷花药膳。（凌弘 摄）

The unique geographical location and natural conditions of Wujaihoushan have created the unique quality of Magnolia. Magnolia and its barks can be used as medicine or table diet, which has the effect of health preservation and disease treatment. The picture shows the magnolia medicinal diet. Photographed by Ling Hong

作为中国农业文化的重要组成部分，四川江油辛夷花传统栽培体系具有丰富的地域历史内涵和地域文化特色，是典型的川西北山地药用植物文化的代表，也是李白文化所承载的地方文脉与传统文化历经沧桑的见证，极具考古价值、旅游价值和开发价值。图为游客如织的吴家后山。（徐卫、吴成惠 摄）

As an important part of Chinese agriculture, the Jiangyou Traditional Magnolia Cultivation System has rich regional historical connotations and regional cultural characteristics. It is a representative of the typical Alpine traditional Chinese Medicine in Northwest Sichuan. It is also a witness of the changes of local memories and traditions loaded with Li Bai culture. It has great archaeological value, value of tourism and development value. The pictures show the tourist-ridden Wujiahoushan. Photographed by Xu Wei and Wu Chenghui

山区稻作梯田的典型代表
云南红河哈尼稻作梯田系统

作为唯一一条发源于云南的国际性河流，红河在云南孕育出了灿烂的农耕文化——在红河南岸的哀牢山区中，有一片面积达100万亩的梯田，覆盖了元阳、红河、金平、绿春四县。这片梯田，由以哈尼族人为主的十多个民族共同修筑完成，被人们称作"红河哈尼梯田"，自开垦之日算起，距今已有1300多年的历史，养育着哈尼族等10多个民族约126万人口。云南中部的哀牢山地区，地跨热带和亚热带，气候类型复杂多样，动植物资源丰富。连绵起伏的高大山脉，阻挡了北来的寒流和西南来的暖湿气流，使其西南部地区降水丰富，气温高于同纬度和海拔的东部地区，同时又少受冬季的自然灾害的影响。在经过漫长的迁徙之后，哈尼族人选择了在这气候适宜的哀牢山下、红河边上定居。他们在海拔700—2000米的山坡上开垦梯田，把村寨建在梯田的上方，形成了"森林在上、村寨居中、梯田在下，水流穿梭其中"的农业生态景观。哈尼梯田最高垂直跨度达1500米，坡度最高达75°，最小的田块仅有1平方米。

耕地的来之不易让哈尼人更加懂得利用土地。他们在梯田上种植水稻；在稻田里养鱼养鸭；在田埂上种植黄豆；在森林里种植林下作物；在荒山坡地种植苞谷、薯类；在房前屋后种植水果、蔬菜；又在河谷地带种植热带经济林果。凡是能进行农业生产的土地，全被哈尼人悉心利用起来。哈尼人对水的利用与分配，更能够凸显出他们的智慧——穿梭在梯田中的大小沟渠便是证明。为了灌溉梯田，哈尼人在梯田中挖凿了大量的水沟，用来接住森林中的流水和山泉水。每个水沟的分叉处放置一个分水木刻，可以准确计量每个子水沟灌溉梯田的用水量，将水资源进行合理分配。灌溉的同时，肥料也可以借助水沟冲到指定的梯田里，这被称作"水沟冲肥"。

哈尼稻作梯田系统具有极高的经济、科学、生态和文化价值。哈尼稻作梯田系统充分利用并遵循自然的劳作传统，创造了哈尼族丰富灿烂的梯田文化，哈尼族以梯田稻作为生，衣食住行、人生礼仪、节日祭典、信仰宗教、生产生活、哲学思想等，无不打上梯田文化的烙印，梯田稻作文化成了哈尼族文化的本根，也集中展现了中华民族天人合一的思想文化内涵。

地理位置：北纬22°26′—24°45′、东经101°47′—104°16′，地处云南省红河哈尼族彝族自治州。
气候特点：亚热带高原季风气候。
认定时间：2010年被联合国粮农组织列为全球重要农业文化遗产，2013年被联合国教科文组织列为世界文化遗产，同年被农业部列为首批中国重要农业文化遗产。

Geographical location: 22°26′-24°45′N, 101°47′-104°16′E; located in Honghe Hani and Yi Autonomous Prefecture, Yunnan Province.
Climate type: subtropical plateau monsoon climate.
Time of identification: In 2010, it was listed as one of the GIAHS by the FAO. In 2013, it was listed as a World Cultural Heritage Site by UNESCO, in the same year, it was listed as one of the first batch of the China-NIAHS by the Ministry of Agriculture.

The Perfect Example of Mountainous Rice Terraces

Honghe Hani Rice Terraces System

As the only international river originating in Yunnan, Red River has cultivated a splendid farming culture in Yunnan. In the Ailao Mountain area on the south bank of the Red River, there is a terraces system of 1 million mu, covering four counties of Yuanyang, Honghe, Jinping and Green Spring. This terraced land was built by more than a dozen ethnic groups dominated by the Hani people. It is known as the "Honghe Hani Terraces". It has been more than 1,300 years since its reclamation, raising a population of 1.26 million from Hani and other ten minorities.

Ailao Mountains in central Yunnan Province, spanning the tropics and subtropics, have complex and diverse climatic types and abundant animal & plant resources. The rolling mountains block the air current from the north and the warm and wet air current from the southwest, making the southwest region rich in rainfall and higher in temperature than the eastern region of the same latitude and elevation, and less affected by natural disasters in winter.

After a long migration, the Hani people chose to settle down at the foot of Ailao Mountain and by the Red River, where the climate is suitable. They reclaimed terraced fields on hillsides between 700 and 2000 meters above sea level and built villages on top of terraced fields, forming an agroecological landscape of "forests on top, villages in the middle, terraced fields at the bottom, water flowing through them". The highest vertical span of Hani terraces is 1500 meters, the highest slope is 75 degrees, and the smallest field is only 1 square meter.

The hard-earned arable land makes the Hani people more aware of the use of land. They grow rice on terraces, fish and ducks in paddy fields, soybeans on ridges, grow crops in forests, grain and potatoes on barren hillsides, fruits and vegetables in front and behind houses, and tropical economic fruits in valleys. All the land that can be used for agricultural farming has been fully utilized by the Hani people.

Hani people's use and distribution of water can more highlight their wisdom, as evidenced by the large and small ditches shuttling in terraces. In order to irrigate terraces, they dig trenches in the terraces to store running water and spring water in the forest. A water dividing woodcut is placed at the bifurcation of each ditch, which can accurately measure the water consumption of the terraces irrigated by each ditch and distribute the water resources reasonably. At the same time, fertilizers and manures used in farming can also be flushed into designated terraces via ditches, which is called "applying fertilizers through ditches".

The Hani Rice Terraces System has a high economic, scientific, ecological and cultural values. It fully utilizes and follows the natural tradition of labor, creating a rich-and-beautiful terraced culture. For the Hani people, the terraces act as a must in their food, shelter and clothing, life etiquette, festivals, religious beliefs, farming and life, philosophy and so on. All of them are branded with terraced rice culture, which has become the root of Hani culture and shows the ideological and cultural connotations of the unity of nature and man in the Chinese nation.

云南红河哈尼稻作梯田系统是利用山地资源，建设形成的"林—寨—田—河"四度同构生态农业系统，分布于云南省南部红河南岸哀牢山区，是亚热带山岳地区稻作生态农业的杰出范例。（黄兴能 摄）

Distributed in Ailao Mountains on the southern bank of Red River in Yunnan Province, Hani Rice Terraces System is a four-dimension isomorphic eco-agricultural system of "forests-villages-terraces-rivers" by utilizing mountain resources. It is an outstanding example of rice eco-agriculture in the subtropical mountain areas. Photo by Huang Xingneng

梯田的田埂非常重要，起着保持水土的关键作用。田埂是用开挖梯田时挖出来的大土饼垒起的，分上埂和下埂，二者的高度因山势缓陡而有所区别。山峦越高越陡峭，因此越往高处，田埂越要不断地加厚、加高，甚至可达4—5米。图为当地人在田间耕作铲埂。（马理文 摄）

Terrace ridges play a key role in soil and water conservation. The ridge is built with large earth cakes dug out during terrace excavation. The ridges are divided into upper ridges and lower ridges. The heights of the two ridges are different based on whether the hills have gentle or steep slopes. The higher the hills are, the steeper their slopes are. So the higher they go, the thicker and the higher the ridges will be, even up to 4-5 meters. The picture shows the local people plowing and shoveling ridges in the fields. Photographed by Ma Liwen

哈尼梯田主要分布在红壤、黄壤和紫色土的分布区内，其土母质基本上是板岩、砂岩、页岩和花岗岩等风化形成的产物，加之生物作用强烈，土质黏重，易保水保肥，如此的土质条件有利于梯田的开垦。同时，哈尼人年年都要修一次田埂，维修时用水稻土打田埂，更增加了田埂的保水保土功能。图为当地人在田间耕作打埂。（马理文 摄）

Hani terraces are mainly distributed where red soil, yellow soil and purple soil occur. The parent materials of soil formation are basically the weathering products of slate, sandstone, shale and granite. In addition, they have strong biological activity, heavy clay and can save water and fertilizer well. Soil conditions are conducive to the reclamation of terraces. Meanwhile, Hani people have to repair ridges once a year. While doing so, they use paddy soils to press ridges tightly, which increases the function of the ridges in conserving water and soil. The picture shows the local people are plowing fields and creating ridges. Photographed by Ma Liwen

西南地区・Southwest China/云南省・Yunnan Province

板田即土壤板结的田，收获庄稼之后或者开春之时都要将板结的田深翻。因地理原因，很多地方的梯田的板田还是需要牛耕，且需要犁得够深，每处都要犁到，这需要讲究耕作技术。图为田间耕作——犁板田。（马理文 摄）

Hardened field is one with compacted soil. After harvesting crops or in the spring, it is necessary to turn the compacted field deeply. Geographically, cattle are still needed to plow the hardened fields in many places. And it is necessary to plow deep enough and everywhere. This needs farmers to pay attention to plowing techniques. The picture shows field tillage-plowing compacted field. Photographed by Ma Liwen

在长期的农业生产中，哈尼人选育出大量适合当地环境的优良稻种。哈尼梯田的水稻品种因梯田海拔高度、土质等不同而多种多样。不仅如此，不同品种的水稻其栽种深度和密度也不甚统一，需要讲究栽种技术。图为田间耕作——栽秧。（黄兴能 摄）

In the long-term agricultural production, Hani people have bred a large number of excellent rice types right for local environment. In Hani terraces, rice varieties vary with altitude and soil quality. Also, the depth and density of rice planting are not uniform, with a specific planting technique required. The picture shows farming in the field - planting seedlings. Photographed by Huang Xingneng

哈尼人依山造田，传统的森林保护理念使这里的自然生态系统保存良好，为梯田提供着丰富水源。哈尼族创造发明了"木刻分水"和"水沟冲肥"，利用发达的沟渠网络将水源进行合理分配，同时为梯田提供充足肥料。图为收获金秋。（马俊勇 摄）

The Hani people build farmland in the mountains, and their traditional concept of forest protection makes it possible that the natural ecosystem here is well preserved and provides abundant water for terraces. The Hani people have created the farming techniques like Wood Carving Water Separation and Gully Irrigation. In this way, the Hani people use a well-developed network of ditches to rationally distribute water sources while providing sufficient fertilizer for terraces. The picture shows the golden autumn harvest. Photographed by Ma Junyong

哈尼人珍惜土地资源，房前屋后的空地都用来种菜，路边的墙缝也会成为菜地。此外，屋旁箐沟凡是有水的地方就会用来养鱼，鱼在池塘里面，池塘水面养浮萍，浮萍喂猪，猪粪喂鱼，鱼长大后又被放回梯田……图为鱼米丰收的当地农民。（马俊勇 摄）

The Hani people cherish the land resources, and the open space in front of and behind the house is used to grow vegetables, and the wall cracks on the roadside will become vegetable fields. In addition, the ditch water around a house will be used to raise fish. The fish is under the pond while the duckweed grows on the surface of the pond. The duckweed feeds the pig, and in turn the pig manure feeds the fish. When the fish grows up, it will be put back to the terraces ... The picture shows the local farmers who have a good harvest of fish and rice. Photographed by Ma Junyong

哈尼族有一套完整的梯田祭礼体系。从社会功能角度来说，这些活动确保了稻作农耕的绵延传承。图为一年中梯田祭礼之祭水口和祭谷穗。（马俊勇 摄）

Hani people have a complete set of terrace sacrifices. From the perspective of social functions, these rituals ensure the continued rice farming. The pictures show the sacrifice rituals performed to honor water mouths and ears of grain on terraces in a year. Photographed by Ma Junyong

"昂玛突"是哈尼族梯田农耕活动中的重要节日，每年春耕开始前举行，一般持续3-5天。在此期间，人们要祭祀祖宗神灵，以祈求风调雨顺、五谷丰登、人畜兴旺。祭祀当天，人们要到寨神林举行祭祀仪式，并将牲肉分给各户，认为其能消灾避难。（马俊勇 摄）

"Hhaqma Tul" is an important festival in the terraced farming activities of Hani people. It is held every year before the beginning of spring plowing and usually lasts 3-5 days. During this period, people had to offer sacrifices to their ancestors and gods, to pray for good weather, grain and livestock prosperity. On the day of the sacrifice, people will go to the village god forest to hold a sacrifice ceremony, and the meat will be distributed to each household, that it can eliminate disaster and shelter. Photographed by Ma Junyong

部分地区的哈尼人,第二天还要举行盛大隆重的"长街宴",招待亲朋好友,祈求天地赐福百姓,一年一度的农耕活动由此拉开序幕。(马俊勇 摄)

The Hani people in some areas will hold a grand "long street feast" on the second day to entertain friends and family, pray to heaven and earth to bless the people, and the annual farming activity will kick off. Photographed by Ma Junyong

世界茶树原产地和茶马古道起点
云南普洱古茶园与茶文化系统

生活在青藏高原等高寒地区的人们，常年以肉奶为主要食物来源。高热量食物的摄入，经常会给人带来油腻感，而茶叶泡水，能够缓解油腻的不适感。

唐代，饮茶之风在中国兴盛起来。茶叶的饮用和烹煮之法传入西藏后，立即获得了藏民的喜爱。不过，这些高寒地区的气候，并不适宜茶树生长。于是，在过去，进口茶叶就成了藏区的头等大事。为了获得茶叶，西南地区的民间自发形成了贸易互市，藏民用马匹向内陆人换取藏区不易得到的茶叶等物。《云南志略》（成书于元代）中记载，"交易五日一集，以毡、布、茶、盐互相贸易"。久而久之，汉藏之间形成了几条运输货物的贸易通道，这些贸易通道便是如今所说的茶马古道。

明清时期，长期向藏区运输茶叶的通道川藏茶马古道因战乱被破坏。于是，滇藏茶马古道成为向藏区输送茶叶的主要通道。在滇藏茶马古道上，云南产的普洱茶是最主要的大宗产品。

普洱茶是以云南特有的大叶茶种制作的茶类。云南大叶茶主产地在普洱、西双版纳、临沧等地区。其中，普洱府（今云南普洱）的普洱山所产的茶叶性温味香。于是人们用"普洱"来为这种茶叶命名。明代的地方志《滇略》中，"士庶所用，皆普茶也，蒸而团之"，应是关于对普洱茶最早的文字记载。

普洱茶的名字越叫越响。明代时，普洱府已设官职专门管理茶叶交易。清代，普洱茶成为皇室贡茶，普洱府也成了普洱茶生产和贸易的集散地，是茶马古道的起点，也成了茶文化的中心地带，并形成了"普洱—昆明官马大道""普洱—大理—西藏茶马大道"等6条保存完好的茶马古道，被称为"世界上地势最高的文明文化传播古道"。

现在，普洱市的茶园面积有300多万亩，市内分布着40余处共约117万亩野生茶树群落，有树龄2700年的千家寨野生古茶树和古老的人工栽培的千年万亩古茶园，是野生茶树群落和古茶园面积最大、古茶树和野生茶树保存数量最多的地区。云南普洱古茶园与茶文化系统就是以普洱茶及茶文化为核心，包括古茶树资源与古茶园生态系统、相关传统知识及其应用和普洱茶文化的复合系统。

地理位置：北纬22°02′—24°50′、东经99°09′—102°19′，地处云南省普洱市。
气候特点：亚热带季风性湿润气候。
认定时间：2012年被联合国粮农组织列为全球重要农业文化遗产，2013年被农业部列为首批中国重要农业文化遗产。

Geographical location: 22°02′-24°50′N, 99°09′-102°19′E; located in Pu'er City, Yunnan Province.
Climate type: subtropical monsoon humid climate.
Time of identification: In 2012, it was listed as one of the GIAHS by the FAO.In 2013, it was listed as one of the first batch of the China-NIAHS by the Ministry of Agriculture.

The Origin of Tea Planting and Starting Point of Ancient Tea Horse Road
Pu'er Traditional Tea Agrosystem

People living in alpine areas such as the Qinghai-Tibet Plateau always use meat and milk as their main food source. The intake of high-calorie foods often brings greasy feel, while tea soaked in water can relieve oily discomfort.

In Tang Dynasty, tea drinking was popular in China. After tea drinking and cooking were introduced to Tibet, they immediately gained the favor of Tibetans. However, the climate in these alpine areas are not suitable for the growth of tea plants. Therefore, in the past, importing tea has become the top priority in Tibet. In order to obtain tea, the folks in southwest China spontaneously formed trade exchanges. With inland people, Tibetans exchanged horses for tea and other items that were not easily available in Tibet. As recorded in *Chronicles of Yunnan* (written in Yuan Dynasty), there was a farmer's market every five days where felts, cloth, tea and salt were exchanged. Over time, several trade routes have been formed between Han and Tibet, known as the Ancient Tea Horse Road today.

During the Ming and Qing Dynasties, the Ancient Sichuan-Tibet Tea Horse Road, a long-term channel for transporting tea to Tibet was broken. As a result, Yunnan-Tibet Tea Horse Road has become the main force to transport tea to Tibet ever since. Pu'er Tea, a local specialty of Yunnan, is the the most important bulk product on the Ancient Tea Horse Road.

Pu'er Tea is a kind of tea made from Yunnan's unique large-leaf tea, which is mainly produced in Pu'er, Xishuangbanna, Lincang, etc. Tea of Pu'er Mountain in Pu'er Prefecture (today's Pu'er City) is so warm and fragrant that people use "Pu'er" to name it. In the *Local Chronicles of Yunnan* of Ming Dynasty, the earliest written record of Pu'er Tea should be that "Pu'er" tea is drunk by both common folks and scholars and officials. The fresh leaves are braised and pressed into a pie."

The Pu'er tea was becoming increasingly famous. In the Ming Dynasty, Pu'er Prefecture had set up official posts to take charge of tea trading exclusively. In the Qing Dynasty, it even became the Royal tribute tea. Pu'er Prefecture also became a distribution center for its production and trade. The place also acted as the starting point of the Ancient Tea Horse Road and the center of tea culture. Six well-preserved ancient tea horse roads such as "Pu'er-Kunming Tea Horse Road" and "Pu'er-Dali-Tibet Tea Horse Road ", which were formed then, are known as "the highest ancient roads of spreading civilization and culture in the world".

At present, there are more than 3 million mu of tea gardens in Pu'er City. There are more than 40 wild tea tree communities with a total area of about 1.17 million mu in the city. There is a 2,700-year-old wild tea tree in Qianjiazhai, and ten thousand mu of artificially cultivated tea plantations a thousand years old. So this area is home to the largest well-preserved wild tea tree communities and ancient tea plantations, and the most ancient tea trees and preserved ancient wild tea trees. Yunnan Pu'er Traditional Tea Agrosystem is one composed of Pu'er tea and tea culture as the core, including ancient tea tree resources and ancient tea garden ecosystem, and relevant traditional knowledge and application.

普洱市别称"思茅",是云南地级市,位于云南省西南部,是普洱茶的主要产区,下辖思茅区、宁洱哈尼族彝族自治县、墨江哈尼族自治县、景东彝族自治县、景谷傣族彝族自治县、镇沅彝族哈尼族拉祜族自治县、江城哈尼族彝族自治县、孟连傣族拉祜族佤族自治县、澜沧拉祜族自治县、西盟佤族自治县。这里的人、茶叶、茶文化沿着茶马古道向国内外扩散,将普洱茶带出大山,走向世界。左页图为镇沅千家寨2700多年的野生古茶树。(苏云 摄)

Pu'er City, also known as Simao, is a prefecture-level city in Yunnan Province. It is located in the southwestern part of Yunnan Province and is the main producing area of Pu'er tea. It consists of Simao District, Ning'er Hani and Yi Autonomous County, Mojiang Hani Autonomous County, Jingdong Yi Autonomous County, Jinggu Dai and Yi Autonomous county, Zhenyuan Yi-Hani-Lahu Autonomous County, Jiangcheng Hani and Yi Autonomous County, Menglian Dai- Lahu -Wa Autonomous County, Lancang Lahu Autonomous County, Ximeng Wa Autonomous County. The people, tea and tea culture here spread along the Ancient Tea Horse Road to home and abroad, bringing Pu'er tea out of the mountains and going to the world. The picture on the left page shows the ancient wild tea tree over 2,700 years old in Qianjiazhai. Photographed by Su Yun

景迈千年万亩古茶园位于澜沧拉祜族自治县东南部惠民乡景迈村和芒景村,平均海拔1400米,年平均气温18℃。古茶树分布范围包括景迈、芒景、芒洪、翁居、翁洼等地,总面积2.8万亩,如今有成林成片的采摘面积1万余亩,系当地布朗族、傣族先民所驯化、栽培,迄今已有1000余年的历史,故称"景迈千年万亩古茶园"。(贾宽明 摄)

Jingmai Millennium Million Mu Ancient Tea Garden is located in Jingmai Village and Mangjing Village, Huimin Township, southeast of the Lancang Lahu Autonomous County. Its average elevation is 1400 meters and the annual average temperature is 18 °C. The ancient tea trees distribute among villages such as Jingmai, Mangjing, Manghong, Wengju, Wengwa and others. It covers a total area of 28,000 mu, of which the current large harvesting area is more than 10,000 mu. These tea trees were originally domesticated and cultivated by the local Bulang and Yi ancestors more than 1,000 years ago. That's why it is called "Jingmai Millennium Million Mu Ancient Tea Garden". Photographed by Jia Kuanming

茶马古道源于古代西南边疆的茶马互市，兴于唐宋，盛于明清，抗战中后期最为兴盛。茶马古道分川藏、滇藏两路，连接川滇藏，延伸入不丹、锡金、尼泊尔、印度境内，直到西亚、西非红海海岸。滇藏茶马古道南起云南茶叶主产区思茅，中间经过今天的大理白族自治州和丽江地区、香格里拉进入西藏，直达拉萨。茶马古道是古代中国与南亚地区一条重要的贸易通道，而普洱是茶马古道上独具优势的货物产地和中转集散地，具有悠久的历史。图为茶马古道上的雕塑和曾经的马蹄印。（贾宽明 摄）

The Ancient Tea Horse Road originated from the tea-and-horse trade in the ancient southwestern frontier. It flourished in the Tang and Song Dynasties and flourished in the Ming and Qing Dynasties, and flourished most in the middle and late period of the Anti-Japanese War. The Ancient Tea Horse Road consists of Sichuan-Tibet and Yunnan-Tibet routes, linking Sichuan-Yunnan-Tibet and extending into Bhutan, Sikkim, Nepal and India until the Red Sea coast of West Asia and West Africa. From the south of Ancient Yunnan-Tibet Tea Horse Road, Simao, the main tea producing area of Yunnan Province, entered Tibet through today's Dali Bai Autonomous Prefecture, Lijiang Region and Shangri-La, and went directly to Lhasa. The Ancient Tea Horse Road is an important trade channel between ancient China and South Asia, and Pu'er is a unique place of origin and transit distribution of goods on the Ancient Tea Horse Road with a long history. The pictures show the sculpture on the Ancient Tea Horse Road and the former horseshoe prints in the past. Photographed by Jia Kuanming

和其他作物一样，在人工栽培以前，普洱的茶叶生产经历了一个野生采集、驯化、野生茶树的自然演变或引入种植、形成今天的栽培型茶树品种现代生态茶园等阶段。栽培型茶树经过人工多代驯化以后品系相对稳定，制成品已无毒性，适宜人们饮用。普洱市现存仍在利用的古茶园多以栽培型茶树为主，它们中最年轻的也已有100多岁。图为栽培型普洱茶的叶、芽、花、籽。（贾宽明 摄）

Like other crops, Pu'er tea trees experienced a growth process of wild collection, domestication and natural evolution or introduction of planting, formation of today's cultivated tea varieties and modern ecological tea gardens. The cultivated tea tree is relatively stable after being artificially multi-generation domesticated, and the finished product is non-toxic and suitable for people to drink. The ancient tea gardens still in use in Pu'er City are mainly cultivated tea trees, and the youngest of them are over 100 years old. The picture shows the leaves, buds, flowers and seeds of cultivated Pu'er tea. Photographed by Jia Kuanming

以普洱市为中心的澜沧江中下游少数民族悠久的种茶、制茶历史孕育了独具风格的民族茶道、茶艺、茶俗等内涵丰富的茶文化和饮茶习俗。不同民族对茶的加工和饮用方式更是各具特色。如傣族的"竹筒茶"、哈尼族的"土锅茶"、布朗族的"青竹茶"和"酸茶"、佤族的"烧茶"、拉祜族的"烤茶"、彝族的"土罐茶"等已作为传统的饮茶习俗，代代相传。图为哈尼族祭祀古茶树和佤族饮茶习俗。（苏云 摄）

The long history of tea cultivation and tea-making of the minority nationalities in the middle and lower reaches of the Lancang River, centered on Pu'er City, has nurtured rich tea cultures and tea-drinking customs with unique styles, such as national tea ceremony, tea art and tea custom. Different ethnic groups have their own characteristics in processing and drinking tea, such as "bamboo tube tea" of Dai nationality, "earth pot tea" of Hani nationality, "green bamboo tea" and "sour tea" of Bulang nationality, "baked tea" of Wa nationality, "baked tea" of Lahu nationality and "earth pot tea" of Yi nationality. The diverse customs have been passed down from generation to generation as traditional tea drinking customs. The pictures show the Hani people offering sacrifices to ancient tea trees and Wa people drinking tea. Photographed by Su Yun

在长久的制茶过程中，普洱茶也形成了独特的工艺，杀青、揉捻、晒干、压制成形的技艺由来已久。传统普洱茶是以云南大叶种晒青毛茶直接蒸压而成，大多为团、饼、砖、碗臼等外形，在茶马古道漫长的运输途中再逐渐发酵。图为普洱贡茶与普洱七子饼贡茶。（苏云 摄）

In the process of making tea, Pu'er tea farmers have long developed unique techniques, including tea green removing, rolling, sundrying and pressing. Traditional Pu'er tea is one made of Yunnan big leaf tea species by being directly braised and pressed, mostly in the form of dough, cake, brick and bowl. Then it is gradually fermented during the long transportation on the Ancient Tea Horse Road. The pictures show Pu'er tribute tea and Pu'er seven-cake tribute tea. Photographed by Su Yun

普洱茶制作的第一道工序是对鲜叶杀青。普洱茶选择云南大叶种茶为原料,杀青方法通常为锅炒杀青。由于大叶种茶水分含量高,晾晒往往难以达到杀透的效果,因而多用锅,采用焖抖结合的方式使茶叶迅速失水,达到杀透杀均的目的。图为普洱茶杀青。(贾宽明 摄)

The first step in making Pu'er tea is to remove water of the fresh leaves. Yunnan big leaf tea is chosen as the raw material, which is usually stir-fried in the pot. Due to the high moisture content of the large leaf tea, it is often difficult to remove water thoroughly by sundrying. Therefore, it is necessary to use pots to quickly lose water by combining braising and shaking, so as to achieve the goal of removing all the tea water thoroughly and evenly. The picture shows the removing water of Pu'er tea. Photographed by Jia Kuanming

普洱茶与当地美食相结合，成为极具特色的普洱美食。（贾宽明 摄）

Pu'er tea with local cuisine makes a special Pu'er food. Photographed by Jia Kuanming

为了有效保护"普洱古茶园与茶文化系统",近年来普洱市人民政府制定了一系列条例,采取了一系列措施,加强对普洱茶原产地的保护,树立普洱茶品牌,不断提高质量、优化品质,提供更好的生态、绿色、安全的产品。图为普洱茶艺大赛。(贾宽明 摄)

In recent years, aiming to better protect the "Pu'er Traditional Tea Agrosystem", the People's Government of Pu'er City has proposed regulations and measures to improve the protection of Pu'er tea origin, create a Pu'er tea brand, continuously improve quality, optimize quality and provide better organic, green, and safe tea products. The picture shows Pu'er Tea Art Competition. Photographed by Jia Kuanming

1993年，第一届中国普洱茶叶节在普洱举行，与茶文化相关的丰富节目内容和茶叶交易吸引了海内外的众多茶商和茶文化工作者参与。迄今为止，中国普洱茶节已经走过了十多个春秋，成为一个具有国际性、开放性、公益性的茶界盛会。图为茶节祭祖。（贾宽明 摄）

In 1993, the first China Pu'er Tea Festival was held in Pu'er City. The tea-based variety programs and trade attracted merchants and tea culture workers at home and abroad. Up to now, China Pu'er Tea Festival, held more than ten times, has become an international, open and public welfare event in the tea industry. The picture shows ancestor worship at the Tea Festival. Photographed by Jia Kuanming

传统核桃与农作物套作农耕模式

云南漾濞核桃—作物复合系统

在"以形补形"理论的影响下,人们认为食用外观形状像大脑的核桃仁,能够补充大脑的机能。因此,核桃这种坚果在中国十分受欢迎。从东到西、从南到北,从辽宁丹东到新疆塔什库尔干、从云南勐腊到新疆博乐,都能见到人工栽培核桃树的踪影。同时,核桃树因对土壤的适应性强、抗病能力强的特点,也使它成为中国颇受欢迎的经济树木之一。

人们公认核桃这一物种的价值是在汉代时,由张骞从西域带入中原。西晋时期张华写作的《博物志》就是证明:"张骞使西域还,乃得胡桃种。"胡桃便是现在的核桃,也称"羌桃"。在这一观点的影响下,人们普遍认为,中国的核桃是进口品种。但是,云南漾濞出土的一段古核桃木,经科技检测距今已有3500年,比张骞出使西域的时间早了1000多年。

漾濞彝族自治县位于大理白族自治州中部。这里光照充足、雨量充沛,土壤肥沃且保水力强,对于喜光耐涝的核桃来说,是极佳的生长地。漾濞生产的核桃果大、壳薄、仁白、味香。宋代时,在漾濞所属的大理州,即当时的大理国,核桃就作为商品出现在市场贸易中。清代,漾濞核桃的美名已经远扬,地方志《滇海虞衡志》中提到,"核桃以漾濞江为上,壳薄可掐而破之"。

核桃在漾濞是经济林、风景树,还是生态林。这里几乎家家户户都种有核桃树。一颗核桃树从栽种到收获果实,大概需要8—10年的时间。在漫长的等待期,漾濞的农民就在核桃树下种植其他农作物,达到"以短养长"的目的,逐渐形成了核桃与农作物间套作的农耕模式。在核桃树下耕种农作物,还能起到为核桃施肥、松土、除草、浇灌的作用。在这种耕作模式下,核桃生长快、结果早且多,还可以收获其他作物。云南漾濞核桃—作物复合系统即是以林下种养结合为特色的复合生态农业系统。

核桃养育了一代又一代漾濞人。目前,漾濞核桃种植面积达92万亩,年产量2.7万吨,产值已突破5亿元,农民人均核桃纯收入近3000元。

地理位置:北纬25°27′50″—25°36′17″、东经99°51′24″—99°59′50″,地处云南省大理白族自治州漾濞彝族自治县。
气候特点:亚热带和温带高原季风气候。
认定时间:2013年被农业部列为首批中国重要农业文化遗产。

Geographical Location: 25°27′50″-25°36′17″N, 99°51′24″-99°59′50″E; located in Yangbi Yi Autonomous County, Dali Bai Autonomous Prefecture, Yunnan Province.
Climate type: subtropical and temperate plateau monsoon climate.
Time of identification: In 2013, it was listed as one of the first batch of the China-NIAHS by the Ministry of Agriculture.

The Traditional Intercropping Pattern of Walnut and Crops
Yangbi Walnut-Crop Complex System

The theory of "shape-supplement" says each human organ can be nourished by eating food of the organ shape. Based on the theory, people believe that eating brain-shaped walnut kernels can improve the function of the brain. Therefore, walnuts are very popular in China. Whether you travel from east to west like from Dandong in Liaoning Province to Tashkurgan in Xinjiang, or from south to north like from Mengla in Yunnan Province to Bole in Xinjiang, you will always find artificially cultivated walnut trees. At the same time, the walnut tree has strong adaptability to the soil and resistance against diseases, which makes it one of the most popular economic trees in China.

It is generally acknowledged that walnut was introduced from the Western Regions to the Central Plains by Zhang Qian in the Han Dynasty. *The Natural History* written by Zhang Hua in the Western Jin Dynasty proves that when Zhang Qian returned from the Western Regions, he brought back walnut seeds. Under the influence of this view, it is generally believed that walnuts in China are imported varieties. However, a piece of ancient walnut wood, unearthed in Yangbi of Yunnan, has been proven with testing technology to be 3,500 years old, more than 1,000 years earlier than Zhang Qian's trip to the Western Regions.

The Yangbi Yi Autonomous County is located in the middle of the Dali Bai Autonomous Prefecture. It is full of sunshine, abundant rainfall, fertile soil and strong water retention. It is an excellent place for the walnuts that are light resistant. The walnut fruit produced in Yangbi is large in shape, thin in shell, white in color and fragrant in flavor. In the Song Dynasty, in the Dali Prefecture and the Dali Kingdom at that time, walnuts appeared in the market trade as commodities. In the Qing Dynasty, the name of the walnut was a household word. *The local Yuheng chronicles of Yunnan Sea* mentioned that "the walnut of the Yangbi River is of superior quality, and its thin shell can be easily pinched open."

Walnut trees in Yangbi act as economic forests, landscape trees, and ecological forests. They are grown in almost every household here. It takes about 8 to 10 years for a walnut tree to grow fruits. During the long waiting time, the farmers plant other crops under the walnut trees to achieve the goal of "short-term growth" and gradually form a farming model of intercropping between walnuts and crops. Planting crops under the walnut trees can also play a role in fertilizing, loosening, weeding and watering the walnut. In this farming mode, walnuts grow fast, bear early fruit, bear more fruit, and farmers can also harvest other crops. The Yunnan walnut-crop complex system is a typical complex eco-agricultural system featuring planting and cultivation under the forest.

Walnuts have nurtured generations of people. At present, the planting area of Yangbi walnut has reached 920,000 mu, with the annual production capacity of 27,000 tons. And the output value has exceeded 500 million yuan with the per capita net income of growers nearly 3000 yuan.

云南漾濞核桃—作物复合系统遗产地——光明万亩核桃生态园，属大理白族自治州漾濞彝族自治县苍山西镇，涵盖整个光明村，地处苍山腹地，总面积15.73平方千米。漾濞核桃历史源远流长，可追溯到3500多年前。图为夜空下的漾濞光明生态核桃树。（黄兴能 摄）

The core area of the Yangbi Walnut-Crop Complex System Guangming Walnut Eco-Park, belongs to Cangshan West Township, Yangbi Yi Autonomous County, Dali Bai Autonomous Prefecture. It covers the whole Guangming Village and is located in the hinterland of Cangshan Mountain with a total area of 15.73 square kilometers. Yangbi walnut has a long history, which can be traced back to more than 3500 years ago. The picture shows the Guangming Eco-park walnut trees in the night sky. Photographed by Huang Xingneng

光明核桃是漾濞核桃的典型代表，早在公元前16世纪就有核桃生产，现在全村树龄在200年以上的核桃树约有6000多株。上图为光明村的古核桃树，下图中的古核桃树已有1160余年的历史。（上图 涂序波 摄；下图 黄兴能 摄）

Guangming walnut is a typical representative of Yangbi walnut. Now there are more than 6,000 walnut trees over 200 years old in the whole village. The pictures show the ancient walnut trees in Guangming Village. The ancient walnut tree in the picture below has a history of more than 1,160 years. The picture above photographed by Tu Xubo, and below by Huang Xingneng

漾濞核桃以果大、壳薄、仁白、味香、出仁出油率高、营养丰富而誉满中外。每年每家每户都收获大量的核桃。（黄兴能 摄）

Yangbi walnut is famous for its large fruit, thin shell, white kernel, fragrant flavor, high oil yield and rich nutrition. Every household harvests a large amount of walnuts every year. Photographed by Huang Xingneng

核桃与各种农作物间套作形成的独特农耕模式已彰显其魅力，是云南漾濞核桃—作物复合系统的集中体现。核桃与各种农作物间套作复合栽培，形成以核桃为主，与粮食作物、水果、中草药材、蔬菜、畜禽等间套作和复合养殖的生态农业模式，实现了农业生产良性循环和可持续发展。（涂序波 摄）

The unique farming mode of intercropping walnuts and various crops is charming and an expression of the Yangbi Walnut-Crop Complex System. The walnuts are intercropped with various crops, forming an ecological cultivation model of walnuts and food crops, fruits, Chinese herbal medicines, vegetables, livestock and poultry, etc., which realizes recycling and sustainable development of agricultural production. Photographed by Tu Xubo

目前漾濞县委、县政府高度重视漾濞核桃—作物复合系统保护，制定了保护与发展规划和管理办法，通过多种方式使这一重要农业文化遗产散发出浓郁的农耕文化魅力。其中，生态旅游应运而生，表现出村在林中，房在树中，人在景中的人与自然和谐共融的画卷。（涂序波 摄）

At present, the local county government is attaching great importance to developing, managing and protecting the Yangbi Walnut-Crop Complex System. With a wide range of methods, this important agricultural heritage system exudes a strong and refreshing agrarian culture charm. For example, ecotourism has emerged at the moment, displaying a beautiful picture, where villages are in forests, houses are amid trees and humans live in harmony with nature. Photographed by Tu Xubo

云南壮族稻作文化的代表
云南广南八宝稻作生态系统

广南县地处云南省东南部、文山州东北部,滇、桂、黔三省的交界处。从广南八宝镇走出的八宝米,驰名全国,是国家的名贵稻种之一。八宝稻作生态系统就是以八宝贡米栽培为特色的复合生态农业系统,最早可追溯到公元前 1200 年。该系统涵盖八宝、莲城、旧莫等 15 个乡镇,目前总种植面积 15 万亩,核心区在八宝镇。

广南县是典型的亚热带季风气候,降水丰沛、光热充足、雨热同期。八宝镇位于广南县的东南部,海拔 1100—1300 米。八宝镇的地形复杂多样,兼具了山地、盆地、河谷、丘陵等多种地貌。复杂的地形,也使得八宝镇有着冬无严寒、夏无酷暑、雨量充沛、雨热同季、光照充足的气候特点。

广南是云南省的壮族人聚居的地方。壮族是世界上最早栽培水稻的民族之一。根据资料记载,早在人类从渔猎和采集为生过渡到以农为生的新石器时代,生息繁衍在云贵高原上(包括广南)的壮族先民已开始将野生稻驯化为栽培稻。

精于稻作的壮族人,能够精确地把控稻米的种植生产体系,懂得精选土地、人畜耕种和顺时而为。他们在这片区域驯化和培育的稻米,色泽晶莹透亮。稻米煮熟后,饭粒软和,富于黏性,清香可口。根据《广南府志》记载,明代时,八宝镇生产的稻米被选为贡米,"每岁贡百担,专送京都",供皇家食用,并根据地名将稻米命名为"八宝米"。

在漫长的历史长河中,八宝的壮族人民创造了绚丽多姿的稻作文化,形成了一系列富有稻作特色的农耕文化。八宝稻作就像一块活化石,记录了当地壮族社会、经济、文化的历史发展概貌。

地理位置:北纬 23°29′—24°28′17″、东经 104°31′—105°39′,地处云南省文山壮族苗族自治州广南县。
气候特点:亚热带高原季风气候。
认定时间:2014 年被农业部列为第二批中国重要农业文化遗产。

Geographical Location: 23°29′-24°28′17″ N, 104°31′-105°39′ E; located in Guangnan County, Wenshan Zhuang and Miao Autonomous Prefecture, Yunnan Province.
Climate type: subtropical plateau monsoon climate.
Time of identification: In 2014, it was listed as one of the second batch of the China-NIAHS by the Ministry of Agriculture.

A Representative of Rice Culture of Yunnan Zhuang Nationality

Guangnan Babao Rice Farming Ecosystem

Guangnan County is located in the southeast of Yunnan Province, the northeast of Wenshan Prefecture, and at the junction of Yunnan, Guangxi and Guizhou provinces. Babao rice from Babao Township, Guangnan, is well-known throughout the country and one of the country's precious rice varieties. Babao rice farming ecosystem is a complex agricultural system characterized by the cultivation of Babao tribute rice, traced back to 1200 BC. The system covers 15 townships such as Babao, Liancheng and Jiumo. At present, the total planting area is 150,000 mu, and the core area is Babao Township.

Guangnan County is of a typical subtropical monsoon climate with abundant rainfull, light and heat, and simultaneous rain and heat. Babao Township is located in the southeast of Guangnan County, with an altitude of between 1,100 and 1,300 meters. The terrain of Babao Township is complex and diverse, with a variety of landforms such as mountains, basins, river valleys and hills. This complex mountain terrain has a impact on local climate, which is characterized by mild winters, warm summers, abundant rainfall and sunshine, and simultaneous heat and moisture.

Guangnan is a place where the Zhuang people live in Yunnan Province. The Zhuang nationality is one of the first people in the world to cultivate rice. According to the records, far back in the Neolithic Age, when human beings transited from fishing, hunting and collecting to farming, the ancestors of the Zhuang nationality who lived and reproduce on the Yunnan-Guizhou Plateau (including Guangnan) began to domesticate wild rice into cultivated rice.

The Zhuang people who are skilled in rice cultivation can accurately control the rice planting and production system, and know how to select land, cultivate people and animals and do things in time. The rice they domesticated and bred in this area is bright and shining. After the rice is cooked, the rice grains are soft, sticky and delicious. According to *Chronicles of Guangnan*, in the Ming Dynasty, the rice produced in Babao Township was selected as "Tribute Rice", "every year 100 dan rice was delivered exclusively to the capital, Beijing, " for royal consumption, and was named "Babao Rice" according to the local name of place.

In the long history, the Zhuang people of Babao created a splendid rice culture and formed farming cultures with rich rice characteristics. Babao rice cultivation is like a living fossil, recording the historical development of the whole society, economy and culture of the local Zhuang people.

广南八宝稻作生态系统位于云南省东南部，文山壮族苗族自治州东北部，地处滇、桂、黔三省交界。图为贡米田园——文山广南县八宝镇坡现村。（林颂 摄）

Guangnan Babao Rice Farming Ecosystem is located in the southeast of Yunnan Province, northeast of Wenshan Zhuang and Miao Autonomous Prefecture, and at the junction of Yunnan, Guangxi and Guizhou provinces. The picture shows Tribute Rice fields, Poxian village, Babao Township, Guangnan County, Wenshan. Photographed by Lin Song

广南八宝稻谷属籼稻。八宝米为优质大米，色泽晶莹透亮，营养丰富，早在明清时期就被列为"贡米"。八宝镇无可比拟的日照条件，恰到好处的雨露滋润，滋养了自然天成的八宝稻米。当地壮族人民总结出一套从良种培育更新、播种移栽、田间管理、收割储存到精制加工的传统精耕细作的生产技术。图为坡现村的"中国八宝米"引水渠以及地方农民出工、肥田打谷子和收获的场景。（林颂 摄）

Guangnan Babao rice belongs to indica rice. Babao rice is a high-quality rice with bright color and rich nutrition. It was listed as "Tribute Rice" far back in the Ming and Qing Dynasties. The unparalleled sunshine conditions and the right rain help nourish the rice naturally. In addition, the local Zhuang people have summed up a set of traditional intensive farming techniques, like cultivating improved seeds, sowing and transplanting, field management, first storage and refined processing. The pictures show Poxian village's diversion canal used for irrigating Chinese Babao rice, local farmers working, as well as raking felds, threshing and harvesting. Photographed by Lin Song

在漫长的历史长河中，八宝壮民们创造了绚丽多姿的优秀文化，形成了一系列富有特色的农耕文化。这些都是不可复制的活态文化，就像一块活化石，记录了壮族社会、经济、文化的历史发展概貌。图为当地妇女制作糍粑、八宝米沙糕和八宝米粽粑。（林颂 摄）

In the long history, Babao Zhuang people have created a splendid and colorful culture, forming a series of distinctive farming culture. These are non-reproducible living cultures. She is like a living fossil, recording the historical development of the Zhuang Ethnic Group's society, economy and culture. The pictures show the local women making glutinous rice cakes (ciba), glutinous rice sandcakes (Babao shagao) and glutinous rice dumplings (Babao zongba). Photographed by Lin song

3000年水旱轮作的"活化石"
云南剑川稻麦复种系统

在同一块土地上，按照时间顺序，轮换种植不同作物的农业生产方式被称为"轮作"。轮作是在提高效率、在一块土地上尽可能多地获得收益的基础上，一种以地养地、促进土地养分循环、改善土壤结构的农耕方式。水旱轮作则是其中颇有挑战性的一种轮作方式，在中国主要体现为两大粮食主力军水稻和麦子的复种。云南剑川稻麦复种系统就是以水稻和小麦、水稻和大麦一年两熟复种为特色的复合农业生态系统。

大概3000多年前的新石器时期，生活在现云南剑川海门口的原始先民们，就开始了稻麦复种、水旱轮作的耕作方式。在海门口遗址中，曾出土了稻、粟、麦等多种农作物的遗存，证实了稻麦复种、水旱轮作在我国至少有3000年的历史。绵亘不断的横断山脉阻隔不了稻麦复种农业文化的传播发展，一年两熟的稻麦复种仍然是当今剑川县的主要耕作制度，是传统农业生产发展的历史见证和缩影，是农业文化、生物多样性、人与自然和谐发展的典型代表，具有文化、生态、经济等多重价值。

每年的5—6月份，是剑川农民栽种水稻的时节。水稻收获后的10—11月份，则是在同一片土地上播种麦子的日子。来年5—6月份麦收之后，继续种水稻，依次循环。

这种耕作方式的成功延续，很大程度上得益于剑川的自然气候。剑川地区光热资源丰富，河流纵横交错，水网密布，土壤肥沃，农业自然控害能力较强。优越的自然条件是农业发展的助推剂。雨季时，剑川降水丰富，水利条件好，能够满足水稻生长期对水量的要求。在水稻收获后的11月份，受季风气候的影响，此时的剑川云量稀少，日照丰富，气温较高，降水不多，蒸发量大，水田自然变成旱地，解决了许多水田改旱地而带来的耕作技艺上的困难。

历经3000年而不衰的稻麦复种系统，蕴含着丰富的自然农法思想和生态理念。在提高复种指数外，稻麦复种、水旱轮作的耕作方式，还能减轻病虫和草害，促进土壤内部养分循环，对于生态循环、水土保持、水源涵养、气候调节等方面都有着重要的意义。

地理位置：北纬26°12′—26°47′、东经99°33′—100°33′，地处云南省大理白族自治州剑川县。
气候特点：低纬度高原季风气候。
认定时间：2014年被农业部列为第二批中国重要农业文化遗产。

Geographical location: 26°12′-26°47′ N, 99°33′-100°33′E; located in Jianchuan County, Dali Bai Autonomous Prefecture, Yunnan Province.
Climate type: low latitude plateau monsoon climate.
Time of identification: In 2014, it was listed as one of the second batch of the China-NIAHS by the Ministry of Agriculture.

The "Living Fossil" of 3000-year-old Crop Rotation
Jianchuan Rice-Wheat Multiple Cropping System

On the same piece of land, the agricultural production method of rotating different crops in chronological order is called rotation. Rotation is based on improving efficiency and gaining as much as possible on a piece of land. It is a farming method that uses land to raise land, promotes nutrient cycling, and improves soil structure. Paddy-upland rotation is one of the most challenging ways of rotation. In China, it is mainly reflected in the multiple cropping of rice and wheat. The rice-wheat multiple cropping system in Jianchuan, Yunnan Province is a complex agro-ecosystem featuring double cropping of rice and wheat or barley in one year

About 3,000 years ago, during the Neolithic Age, the primitive ancestors who lived at the Haimenkou of Jianchuan, Yunnan Province, began to cultivate rice and wheat in multiple cropping and paddy-upland rotation. In Haimenkou site, the remains of rice, millet, wheat and other crops have been unearthed, confirming that rice-wheat multiple cropping and paddy-upland rotation have a history of at least 3000 years in China. The rolling Hengduan Mountains can't block the spread and development of the agricultural culture. The double cropping of rice and wheat in one year is still the main farming system in Jianchuan County today. It is a witness and example of the development of traditional agricultural production. It is a typical representative agricultural culture, biodiversity, harmony between human and nature. It has multiple values of culture, ecology and economy.

In May and June of each year, it is the season when Jianchuan farmers plant rice. In October and November after rice harvest, it is the time when wheat is sown on the same land. After the wheat harvest in May and June of the following year, the rice will continue to be planted and recycled in turn.

The success of this farming method is largely due to the natural climate of Jianchuan. The Jianchuan area is rich in light and heat resources, the rivers are criss-crossed, the water network is dense, the soil is fertile, and the agricultural natural control ability is strong. Superior natural conditions are a booster for agricultural development. During the rainy season, Jianchuan has abundant precipitation and good water conservancy conditions, which can meet the water demand in the rice growing season. Affected by the monsoon climate in November after rice harvest, Jianchuan had scarce clouds, abundant sunshine, high temperature, little precipitation and large evaporation, and paddy fields naturally became dry lands, which will solve the problems of farming skills brought about by changing paddy fields into dry lands.

The rice-wheat multiple cropping system, which has lasted for 3000 years, contains rich ideas of natural agricultural law and ecological concepts. In addition to improving the multiple cropping index, rice-wheat multiple cropping, paddy upland rotation and farming methods can also reduce diseases, insects and weeds, promote nutrient cycling within the soil, which has important significance for ecological cycle, soil and water conservation, climate regulation and other aspects.

云南剑川县稻麦复种系统位于云南省大理白族自治州剑川县，涵盖全县 7 万亩水稻面积，核心区为金华镇、甸南镇和沙溪镇，核心区面积 3 万亩。图为剑川甸南镇兴水村的水稻。（苏金泉 摄）

The rice-wheat multiple cropping system in Jianchuan County, Dali Bai Autonomous Prefecture, Yunnan Province, covers 70,000 mu of rice area. The core areas are Jinhua Town, Diannan Town and Shaxi Town, with a total area of 30,000 mu. The picture shows rice in Xingshui Village, Diannan Town, Jianchuan County. Photographed by Su Jinquan

剑川素有"文献名邦""木雕之乡""白族文化的聚宝盆"和"云南文明的发源地"之美誉。每年5—6月份栽种水稻，10—11月份水稻收获后，翻耕播种大麦或者小麦，来年5—6月收获，麦茬翻耕后再栽水稻。水旱轮作，提高复种指数，减轻病虫草害，改善土壤结构，促进养分循环。图为剑川甸南桃源村民在插秧和种植小麦的场景。（苏金泉 摄）

Jianchuan has the reputation of "the famous county in documents", "the hometown of woodcarving", "the cornucopia of Bai culture" and "the origin of Yunnan civilization". Rice is planted from May to June every year. After the rice harvest in October and November, the barley or wheat is ploughed and sowed, and harvested from May to June of the following year. After wheat stubble is plowed, rice is planted again. Paddy-upland rotation is helpful in increasing the multiple cropping index, reducing pests and diseases, improving soil structure, and promoting nutrient cycling. The pictures show a scene of seedling transplanting and wheat planting by people in Taoyuan Village, Diantan Town,Jianchuan County. Photographed by Su Jinquan

自 3000 多年前新石器时代晚期开始，剑川稻麦复种水旱轮作的耕作方式一直沿用至今。春耕夏耘、秋收冬藏，亘古不衰的稻麦复种系统蕴含的生态价值理念、自然农法思想以及古老农具、农耕技术，处处或隐或显地展现了白族先民的身影和智慧。图中村民在做春季秧田的准备。（杨士斌 摄）

Since the late Neolithic Age more than 3000 years ago, the rice-wheat multiple cropping system in Jianchuan has been used till now. Spring ploughing, summer ploughing, autumn harvesting and winter harvesting, the ecological value concept, natural agricultural law thought and ancient farming tools and farming technology contained in the ever-lasting rice-wheat multiple cropping system, show the image and wisdom of Bai ancestors everywhere or implicitly or explicitly. The farmers in the picture are preparing for spring seedling fields. Photographed by Yang Shibin

剑湖是云南重要的高原湿地之一，位于滇西北横断山脉中南段，大理州剑川县境内，紧靠县城。图为剑湖东岸的种植小麦场景和同一地点拍摄的种植水稻场景。（苏金泉 摄）

Jian Lake is one of the important plateau wetlands in Yunnan. It is located in the middle and south section of the Hengduan Mountains in northwestern Yunnan. And it is in the territory of Jianchuan County, Dali Prefecture, next to the county town. The pictures show the wheat planting on the east bank of Jian Lake and the rice planting at the same spot. Photographed by Su Jinquan

继 1957 和 1978 年之后，国家文物局于 2007 年 12 月批准对剑川海门口遗址进行第三次考古发掘。此次发掘共出土编号小件 2000 多件，其中农作物包括有炭化稻、麦、粟、稗子等。图为剑川的水稻和小麦。（苏金泉 摄）

Following the archaeological excavations of 1957 and 1978, the State Administration of Cultural Relics approved the third one of the Haimenkou site in Jianchuan in December 2007. More than 2000 small numbered pieces were unearthed, including carbonized rice, wheat, millet and barnyard grass. The picture shows rice and wheat in Jianchuan. Photographed by Su Jinquan

剑川素有"白族文化聚宝盆"的美誉。当地白族既传承着中华民族的主要岁时节庆,也坚守着丰富多样的民族节俗,其中许多节庆民俗都与稻麦复种农耕文化密切相关。图为稻田里的剑川白族百姓。(赵渝 摄)

Jianchuan is known as the treasure basin of Bai culture. Local Bai people not only celebrate the main festivals of the Chinese nation, but also adhere to rich and diverse ethnic festivals, many of which are closely related to rice-wheat multiple cropping farming culture. The picture shows the Bai people in Jianchuan in rice fields. Photographed by Zhao Yu

传统稻鱼鸭共生农业生产模式
贵州从江侗乡稻鱼鸭系统

对自然规律的利用从来都能体现着人类的智慧。贵州省从江县的劳动人民，利用物候特点对水稻、田鱼和鸭子进行共同培育和养殖便是一例。

从江县地处贵州省东南部的群山之中。这里山连着山，岭接着岭，是云贵高原的边缘地带，地势复杂，层峦叠嶂、沟谷纵横。亚热带季风气候定时给这里带来充沛的降水，都柳江奔腾着自北向南穿境而过，丰富的水资源，让人们对土地的利用增加了更多可能。

每年清明、谷雨前后，居住在从江的人们开始了一年的忙碌，这便是插秧的时节。

农民把水稻秧苗栽种到水田里，同时也把鱼苗放入水田中。这种鱼是在不断的发展中，经过从江特有环境的自然选择以及劳动人民的人工培育所产出的、从江特有的稻田鲤鱼品种。

等到水稻秧苗返青、鱼儿长到三指宽的长短时，幼鸭也被投放到稻田里饲养。此时鸭子尚小，还不能对水稻和田鱼构成威胁。稻田里鸭子的投放，十分讲究。作为杂食动物，如果投放时机不对，鸭子对水稻的生存将构成威胁。从江人民大概在3个时期会把鸭子放入稻田中。除了幼鸭时期，也是水田空置时，刚出生的雏鸭会被放到水田里养到农历三月初水稻播种之时；待到水稻郁闭、田鱼长到8厘米左右时，稻田里又可以放养成鸭。鸭子是从江人民选育出来的小香鸭，个头小，能够在稻田里灵活自如穿行而不撞坏水稻。农民不需特意为鸭子的食物费心，稻田里的昆虫和杂草，便是鸭子的食物。

在这一生产模式中，稻田为鱼和鸭子的生长提供了环境和食物；鱼和鸭子为田里的水稻清理了杂草和害虫；同时，鱼鸭在稻田中的活动，无形中给稻田松了土，而它们的粪便又可以为稻田做肥料。对物候特点的把握和对生物习性的了解，使从江人民在水稻收割的季节，总能收获稻米、鲤鱼和鸭子三种产品。

"种植一季稻、放养一批鱼、饲养一批鸭"是从江侗乡劳动人民世代传承的农业生产方式，以其为特色而形成的稻鱼鸭共生、鱼米鸭同收的复合生态农业系统，正是贵州从江侗乡稻鱼鸭系统的特点。侗乡人利用他们的智慧，将三种原本相克的物种，和谐地编织进了一个生态系统中，并最终达到了互利共生的效果。

地理位置：北纬25°16′—26°05′、东经108°05′—109°12′，地处贵州省黔东南苗族侗族自治州从江县。

气候特点：亚热带季风气候。

认定时间：2011年被联合国粮农组织列为全球重要农业文化遗产，2013年被农业部列为首批中国重要农业文化遗产。

Geographical Location: 25°16′-26°05′N, 108°05′-109°12′E; located in Congjiang County, Southeast Guizhou Miao and Dong Autonomous Prefecture, Guizhou Province.
Climate type: subtropical monsoon climate.
Time of identification: In 2011, it was listed as one of the GIAHS by the FAO. In 2013, it was listed as one of the first batch of the China-NIAHS by the Ministry of Agriculture.

Traditional Rice-fish-duck Symbiotic Agricultural Production Mode
Congjiang Dong's Rice-Fish-Duck System

The use of the laws of nature can always reflect the wisdom of mankind. The working people in Congjiang County, Guizhou Province, use phenological characteristics to jointly cultivate and breed rice, field fish and ducks.

Congjiang County, the marginal region of Yunnan-Guizhou Plateau, is located in the mountains in the southeast of Guizhou Province. Mountains are connected to and followed by the mountains. The terrain is complex, with layers upon layers of peaks and knolls and valleys. The subtropical monsoon climate regularly brings abundant precipitation to the area. The Duliu River runs from north to south through the boundary, and the abundant water resources offer local people with more possibilities of land usage.

Every year, around the Solar terms of Ching Ming and Gu Yu, people living in Congjiang will start their busy year with rice shoot transplanting.

The farmers plant the rice seedlings into the paddy fields together with the fish larvae. Fish is constantly evolving and developing. Through natural selection and special cultivation by the working people, this specific carp is chosen from the characteristic natural environment in Congjiang area.

When the rice seedlings return green and the fish grow up to around 5 - 6cm long, the young ducklings are also placed in the rice fields. At this time, the ducks are still small, and they cannot pose a threat to rice plants and field fish. The placement of ducks in the rice fields is very essential. As an omnivorous animal, ducks could pose a threat to rice growth if the placement timing is not right. The people of Congjiang will put the ducks into the rice fields in about three different periods. Other than the young ducks, when the paddy fields are vacant, the newborn ducklings will be placed in the paddy fields till the beginning of the third lunar month before the transplanting of rice seedlings. When the paddy field is at crown closure, the fish will grow up to about 8 cm, adult ducks can be stocked in the rice fields. This Small Fragrant Duck species is selected by the people of Congjiang. It is small and can travel freely through the rice fields without damaging the rice plant. Farmers don't need to worry about the food of the ducks. The insects and weeds in the rice fields are just the good feeds for the ducks.

In this model of production, rice fields provide the environment and food for the growth of fish and ducks; fish and ducks clean up weeds and pests for the rice in the fields; at the same time, the activities of fish and ducks in the rice fields can invisibly loosen the dirt and soils, and their feces can be used as fertilizer for rice fields. By mastering the characteristics of phenology and understanding of biological habits, the people of Congjiang are able to harvest rice, carps and ducks simultaneously during the harvesting season.

"Growing a season of rice, stocking a batch of fish, and raising a batch of ducks" is an agricultural production method that has been passed down from generation to generation by the working people of Congjiang. It is a combination of symbiosis breeding and harvesting of rice, fish, and ducks. This farming mode is the characteristic of Congjiang Dong's Rice-Fish-Duck System. Using their wisdom, the local farmers harmoniously weaved the three original species into an ecosystem, and finally achieved the effect of mutual benefits.

西南地区・Southwest China/ 贵州省・Guizhou Province

从江县位于黔东南苗族侗族自治州层峦叠嶂的大山里，清澈的都柳江从北向南蜿蜒而过。从江农民稻田养鱼历史悠久，已有1000多年的历史，这里现有稻田17万亩。图为从江加榜梯田三家寨景点，加榜梯田是国家湿地公园。（张琪 摄）

Congjiang County, located in the layers of layers of mountains and knolls in southeastern Guizhou's Miao and Dong Autonomous Prefecture, with the clear Duliu River runs through from north to south. Congjiang farmers have a history of fish and duck farming in the paddy fields for more than 1,000 years, and presently there are about 170,000 mu of paddy fields here. The picture shows the Sanjiazhai scenic spot in Congjiang Jiabang terraced rice field, which is a national wetland park. Photographed by Zhang Qi

从江稻鱼鸭系统是典型的人工生态系统，这种以鸭食虫、以鱼食草、用鸭鱼粪作肥料的稻鱼鸭复合系统，具有保护生物多样性、控制病虫害、调节气候、涵养水源等多种生态功能。图片分别为岜扒村的村民在田里放养小香鸭、岜扒村民在田边数有多少条稻田鱼和从江侗乡稻鱼鸭共生的景象。（张琪 摄）

Congjiang Dong's Rice-Fish-Duck System is a typical artificial ecosystem. This symbiotic system, in which ducks eat insects, fish eats weed, and their manures are used as fertilizers of rice plants, functions well in biodiversity protection, pests and diseases control, climate regulation, and safeguarding water sources, etc. The pictures show that the villagers of Bapa village are stocking small ducks in the fields; the villagers are counting the numbers of the fish in the rice fields and the symbiosis of the fish and ducks in Congjiang Dong area. Photographed by Zhang Qi

侗乡人所养育的鸭种不是一般的鸭种，而是经过世代选育驯化的小香鸭。小香鸭个头小，可以灵活地在水稻间穿行而不会撞坏水稻，而且不用投入精饲料，水稻中的害虫、小虾、小鱼、各种杂草都是上好饵料。侗乡人利用智慧，使稻、鱼、鸭三者和谐共处，互惠互利。图为小香鸭在稻田中嬉戏和稻田鱼（鲤鱼和鲫鱼为主）加工场景。（张琪 摄）

The duck species raised by the Dong people are not ordinary duck species, but small ducks that have been domesticated and selected through generations. The Small Fragrant Duck can flexibly move through the rice without damaging the rice, fine feeds are not needed. The pests, shrimps, small fish, and various weeds in the paddy field are good feeds. The Dong people use wisdom to make the rice, fish and ducks live in harmony and benefit mutually. The pictures show the small fragrant ducks playing in the rice fields, and the processing of the paddy fish (mainly carps and crucian carps). Photographed by Zhang Qi

在从江侗乡稻鱼鸭复合系统中，稻占据着重要的位置。从江县的稻田主要为山冲田和山梁梯田。这里因山体和森林的荫蔽，雾多，湿度大，日照短，昼夜温差大，且多引山泉水灌溉，水温低。在这种条件下，侗乡人种植的糯稻不但能够生长和丰收，还能为鱼和鸭提供丰富的饵料。图为稻田劳作摘禾。（张琪 摄）

In the compound system of rice, fish and duck in Congjiang Dong ethnic area, rice production occupies an important position. The paddy fields in Congjiang County are mainly alluvial fields and ridge terraced fields. Due to the shade of mountains and forests, there are many foggy days of high humidity and short sunshine hours, large temperature difference between day and night, and paddy fields are mostly irrigated with cold mountain spring water. Under these conditions, the people of Dong can still grow the glutinous rice well with good yield, and at the same time, providing abundant feeds for both the fish and ducks. The picture shows the field work of picking the rice. Photographed by Zhang Qi

稻鱼鸭系统不仅是侗族人赖以生存的生计模式，也是侗族传统文化的传承载体。无论是侗族建筑的布局和造型，还是特有的民俗风情，或者传统农业背后的社会组织制度，都与侗族这种传统的生计模式有着密切的关系。图为从江县高增风雨桥和占里村秋收时禾晾遍布全村（禾晾即侗乡人为了晾晒摘下的禾把，可使收获的水稻免受家禽、家畜和鼠类糟蹋而专门在当阳通风处建造的木结构建筑）。（张琪 摄）

The rice-fish-duck system is not only a livelihood model for the Dong people to survive, but also a carrier of the traditional culture of the Dong people. Whether it is the layout and shape of the Dong architecture, or the unique folk customs, or the social organization system behind traditional agriculture, it has a close relationship with the traditional livelihood model of the Dong people. The pictures show the Gaozeng Shelter Bridge in Congjiang County and the autumn harvest in Zhanli Village, the Grain Drying Facilities are scattered throughout the village (a wooden structure built in the ventilated area under sunshine by the Dong people, used to dry the harvested rice and protect it from poultry, livestock and rodents). Photographed by Zhang Qi

从江素有"百节之乡"的美誉。2003 年，从江举办了首届侗族大歌节，此后大歌节历年举行。大歌节期间，从江各旅游景区和重要民族村寨都开展丰富多彩的民俗体验活动。侗族人民素有"歌养心、饭养身"之说，侗族许多优秀的文化、生活习俗、社交礼仪等都靠优美的歌声代代相传。图为 2015 年 3 月贵州从江小黄侗寨的千人侗族大歌活动。（张琪 摄）

Congjiang enjoys a reputation as the Hometown of Hundred Festivals. In 2003, the first Dong Grand Song Festival was held in Congjiang, and it has been held over the years ever since. During the Grand Song Festival, a variety of folk experience activities were carried out from various tourist attractions and important ethnic villages in Congjiang. The Dong people are known for their saying: "songs to nourish their hearts and food to nourish their bodies." Many outstanding cultures, customs, and social etiquettes of the Dong people are passed down from generation to generation by beautiful songs. The picture shows thousands of Dong people chorus in the Xiaohuang Village in Congjiang County of Guizhou in March 2015. Photographed by Zhang Qi

西北地区 · Northwest China

陕西省

干旱地区山地高效农林生产体系——陕西佳县古枣园

甘肃省

古梨树存量最多的梨树栽培体系——甘肃皋兰什川古梨园

农、林、牧循环复合生产体系——甘肃迭部扎尕那农林牧复合系统

"千年药乡"的杰出代表——甘肃岷县当归种植系统

宁夏回族自治区

独特环境、独特品种、独特技艺——宁夏灵武长枣种植系统

新疆维吾尔自治区

大型地下农业水利灌溉工程——新疆吐鲁番坎儿井农业系统

哈密地域文化与财富的标志——新疆哈密市哈密瓜栽培与贡瓜文化系统

干旱地区山地高效农林生产体系

陕西佳县古枣园

佳县位于陕西省东北部的榆林市，依黄河而建县。这里东邻黄河，西南面对黄土丘陵，北对毛乌素沙漠，土石山区、黄土沟壑、沙地几乎是佳县土地的全部形态。因深处内陆，典型的大陆性气候让佳县"十年九旱"，年均降水量不足400毫米。这样的生态条件，看起来并不适合长时间进行农业生产，不过，有一种植物"救活"了佳县，那就是枣树。

枣树耐旱、耐贫瘠，生长在沙地对于它们来说，根本不是什么难题。根系发达的枣树，树根能向四面八方延展生长，高大的树冠如伞盖一般。因此，在佳县，枣树不仅可以是经济果木，还可以发挥水土保持、水源涵养、防风固沙等多种生态作用。

陕西佳县古枣园，是以黄土高原景观及"铁杆庄稼"红枣栽培利用为特色的农业文化系统。3000多年前，居住在这里的先民开始在黄河沿岸的坡地上栽培枣树。他们在漫山遍野的酸枣树中，选取果实味道甜美、更容易存活的树种栽培到黄河滩地中。《齐民要术》对此种选育枣树的方法也有所记载。书中提到，"常选好味者留栽之"，意思就是从酸枣里边挑选那些比较好吃的、更容易栽培的留下来，栽培它。经过千年的发展，枣树的品种已稳定下来，佳县人也早已把枣当作他们的农业主要产业。在佳县的泥河沟村，有一片面积36亩的古枣园，保有各龄古枣树1100余株，树龄千年以上的枣树超过100棵。其中最老的古枣树，树龄有1300多年。

农民给了枣树他们所能提供的最好的生长环境，枣树也毫不吝啬地回馈。它们是佳县的"保命树"，在干旱、霜冻、冰雹等自然灾害时常侵扰的情况下，生长在黄河滩的枣树也能年年挂果、岁岁丰产，被当地人奉为"铁杆庄稼"，养活了这一方水土的人们。独特的地理环境和气候特点，使得佳县红枣有着较高的糖分和特殊的风味，再经过长期的选种育种，现在的佳县红枣个大肉鲜，味道香甜，享誉全国，是中国的名枣之一。佳县有着底蕴深厚的红枣文化历史。每年正月，人们都要敬拜"枣神"，祈求红枣丰收。逢年过节，人们都要制作枣糕、枣馍、枣焖饭等传统食品，以示庆贺。长辈们给孩子吃红枣、戴枣串，希望他们早日长大成人，日子甜甜蜜蜜。久远而又浓郁的红枣文化气息渗透在佳县人的日常生活之中。

地理位置：北纬37°41′47″—38°23′34″、东经110°0′45″—110°45′10″，地处陕西省榆林市佳县。

气候特点：温带大陆性半干旱季风气候。

认定时间：2013年被农业部列为首批中国重要农业文化遗产，2014年被联合国粮农组织列为全球重要农业文化遗产。

Geographical Location: 37°41′47″-38°23′34″N, 110°0′45″-110°45′10″E; located in Jia County, Yulin City, Shaanxi Province.
Climate type: temperate continental semi-arid monsoon.
Time of identification: In 2013, it was listed as one of the first batch of the China-NIAHS by the Ministry of Agriculture. In 2014, it was listed as one of the GIAHS by the FAO.

High-efficiency Agroforestry Production System in Arid Areas
Jiaxian Traditional Chinese Jujube Garden

Jia County is located in Yulin City, northeast of Shaanxi Province, and is built by the Yellow River. It is adjacent to the Yellow River in the east, the loess hills in the southwest, and the Mu Us Desert in the north. The rocky mountain, the loess gully, and the sand are almost the overall look of Jia County. Deep inland, the typical continental climate makes Jia County "9 years of droughts in 10 years", with an average annual precipitation of less than 400 mm. Such ecological conditions do not seem to be suitable for long-term agricultural production. However, there is a kind of plant that "saves" Jia County, which is the jujube tree.

Jujube trees are drought-tolerant and poor-tolerant, and growing in the sand cannot be accounted as a problem for them. The developed roots can grow in all directions; the luxuriantly green crown is like a huge umbrella. Therefore, in the environment of Jia County, jujube trees cannot only be economic fruit trees, but also play various ecological functions such as soil and water conservation, wind protection and sand fixation.

The Jiaxian Traditional Chinese Jujube Garden, Shaanxi Province is an agricultural system featuring the landscape of the Loess Plateau and the cultivation and utilization of "Iron Crops". More than 3,000 years ago, the ancestors who lived here began to cultivate jujube trees on the slopes along the Yellow River. Among the jujube trees in the mountains and plains, they chose and planted trees with sweet and more viable fruits on the Yellow River beach. *Qi Min Yao Shu* also recorded the method of breeding jujube trees. It mentioned that choose the trees with more sweet jujubes and more viable to plant. With the millennium development, the varieties of jujube have been stabilized, and people of Jia County have already regarded jujube as their main agricultural industry. In Nihegou Village of Jia County, there is an ancient jujube orchard of 36 mu, which holds more than 1,100 ancient jujube trees of various ages, and more than 100 jujube trees over 1,000 years old. The oldest jujube tree is more than 1,300 years old.

The farmers gave the jujube the best growing environment they could provide, and the jujubes were not mean to repay their meticulous care. They are the "life-saving trees" of Jia County. In the case of natural disasters such as drought, frost, hail and so on, the jujube trees growing on the Yellow River Beach can also be fruit-bearing and high-yielding every year. They are regarded as "Iron Crops" and "save" the locals. The unique geographical environment and climatic characteristics make Jia red jujubes have high sugar and special flavor. After long-term breeding, the red jujubes now are more delicious and plumper and become one of the most famous jujubes in China. Jia County has a profound history of jujube culture. Every year in the first month of the lunar year, people must worship the "jujube god" and pray for a good harvest. At every festival and at New Year, people will make traditional foods such as jujube cake, jujube strings, jujube millet to celebrate; the elders will give their children red jujubes and jujube strings, hoping they could grow up safely and having a happy life. The rich jujube culture penetrates into the daily life of people of Jia County.

佳县古枣园位于"中国红枣名乡"陕西省东北部榆林市佳县朱家坬镇泥河沟村，是世界上保存最完好、面积最大的千年枣树群。这里现存完整的从野生型酸枣、半栽培型酸枣、栽培型酸枣到栽培枣的驯化过程。泥河沟村也被誉为"天下红枣第一村"。图为秋天的泥河沟村。（贾玥 摄）

The Jiaxian Traditional Chinese Jujube Garden is located in Nihegou Village, "Hometown of Jujubes", in Zhujiawa Township, Jia County, Yulin City, northeast of Shaanxi Province in the northeast of Shaanxi Province. It is the world's best preserved and largest millennial jujube trees group. There is a complete domestication process from wild jujube, semi-cultivated jujube, cultivated wild jujube to cultivated jujube. Nihegou Village is also known as the "first village of the world red jujube". The picture shows the village of Nihegou in the autumn. Photographed by Jia Yue

陕西佳县古枣园是一个包含了枣园管理、枣树栽培、红枣加工、红枣文化和黄土高原特色景观的综合性系统。历史久远的文化与古枣园的栽培和管理方式形成了当地特有的社会组织与文化和知识体系，丰富的民间谚语和传统技术涵盖了枣粮间作、枣园管理以及红枣种植和加工多个方面。图为4月末枣树发芽，乡民们开始在枣林下锄地、起垄，种花生、葱、辣椒等各式作物，以及红枣加工产品。（贾玥、闵庆文 摄）

The Jiaxian Traditional Chinese Jujube Garden is a comprehensive system that includes jujube management, jujube cultivation, jujube processing, jujube culture and the characteristic landscape of the Loess Plateau. The long-standing culture and the cultivation and management of the ancient jujube orchard form a unique local social organization and culture and knowledge system. The rich folk slang and traditional techniques cover the intercropping of jujube and grain, the management of jujube orchard and the cultivation and processing of jujube work in many ways. The pictures show that at the end of April, the people begin their tillage squat and plant peanuts, onions, peppers and other crops in the jujube forest and some processed jujube products. Photographed by Jia Yue and Min Qingwen

枣粮间作，一方面涵养了水土，另一方面肥沃了土壤，有助于枣树和作物的共同生长。图为秋天收枣的同时，还要收获花生、红薯和各类果蔬。而枣熟通常赶上国庆时节，外出工作定居的子女通常会回来与家人团聚连带收枣。（贾玥 摄）

Jujube and grain intercropping, on the one hand, conserves water and soil, on the other hand, fertilizes soil, which contributes to the common growth of jujube trees and crops. The pictures show the harvest of jujube and peanuts, sweet potatoes and various fruits and vegetables. Jujube is usually caught up in the National Day, and the offspring who goes out to work will come back to reunite with their family and help to harvest. Photographed by Jia Yue

泥河沟村不仅守着黄河，而且三面环山。这里早春升温快，枣树发芽相对早，延长了枣的生长期。到了秋季，枣果成熟的时候，这里温差又大，极其有利于糖分和风味的转化，使得泥河沟的油枣成为我国名枣之一。图为黄河滩枣林和泥河沟村民在山崖下晾干鲜枣。（闵庆文、贾玥 摄）

Nihegou Village not only guards the Yellow River, but also is surrounded by the mountains on three sides. Temperature will go up quickly in the early spring, and the jujube germination is relatively early, which prolongs the growth period of the jujube. In the autumn, the large temperature difference here is extremely beneficial to the conversion of sugar and flavor substances, making the Chinese jujubes of Nihegou Village one of the famous jujubes in China. The pictures show the jujube forest in the Yellow River Beach and the fresh jujubes dried under the cliff. Photographed by Min Qingwen and Jia Yue

转九曲又称"转灯"，是从陕北流传出来，并在黄河流域流传着的一种汉族民俗文化活动，也是正月里当地人人参与的一项活动。每到正月十五下午，人们便敲锣打鼓，喜开九曲门。人们用高粱秆栽成一个四方形的图阵，上面再放上用泥做的油灯。"九曲"像个城郭，又似一个迷宫，其回廊没有重复路径。有人把九曲称为"九曲黄河阵"。夜幕降临，360盏油灯同时点亮，锣鼓唢呐齐奏，秧歌队打头进入九曲连环阵。图为设在被列为遗产核心区村古枣林中的黄河九曲阵。（贾玥 摄）

Lamp maze is an ancient custom originated from northern Shaanxi Province and spread in the Yellow River Basin. It is celebrated at afternoon on the fifteenth day of the first lunar month. People beat the drums as the beginning of the ceremony. They use sorghum stalks to make up a square pattern with oil-burning lamps made of mud on it. It likes an enceinte or a maze, and there is no repeating path. Some people also call the Lamp Maze "the Yellow River array". As night falls, the 360 lamp lights up at the same time, and the gongs and drums sing together, and the Yangko team starts to enter the circle. The picture shows the Yellow River array set in the ancient jujube orchard which was list as the core area of jujube heritage. Photographed by Jia Yue

古梨树存量最多的梨树栽培体系
甘肃皋兰什川古梨园

素有"世界第一古梨园"之称的什川古梨园位于甘肃省皋兰县。地处黄土高原的皋兰县。这里山川与沟壑相间，气候干旱，光照时间长，年均蒸发量几乎是降水量的6倍。但什川在生态环境上是皋兰县的一块"飞地"。

什川历史悠久，钟灵毓秀，地灵人杰，是黄河文化孕育的杰作。什川东距兰州约20千米，西距所属皋兰县城20千米，总面积405平方千米，平均海拔1500米，气候湿润，风光秀丽，盛产瓜果蔬菜，是有名的"瓜果之乡"。

什川镇依黄河而建，如孩童般被黄河包揽在怀里。这里三面环水，四周拱山，气候温和，土壤肥沃。明代嘉靖年间（1522—1566年），什川引入梨树，作为经济果木进行栽培。在当地干旱的大环境下，什川农民因地制宜、仿建水车引来黄河水为梨树灌溉。对水量有较高要求的梨树，在什川存活下来。什川独特的地理环境给梨树提供了适宜的生长空间，梨树在这里长势旺盛。此后，梨便成了什川的主要经济作物。甘肃皋兰什川古梨园拥有以数千株树龄200年以上的梨树为特色的复合农业生态系统，遗产地核心区总面积约42平方千米。

如此大面积、连片的古梨园，能够保留到现在，除了得益于什川不易受外界侵扰的地理环境外，还要归功于什川果农世代相传的梨树栽培方法和梨园管理技艺。要为梨树松土、施肥，早春"刮树皮"、花期"堆砂"防虫，更因为梨树树干高大，栽培梨树时需要利用云梯穿梭于半空的梨树间，给果树修枝整形、疏花疏果、竖杆吊枝、采摘果实。什川人将栽培梨树叫作"种高田"。"种高田"所需要的工具云梯被称作"天把式"。"天把式"与树冠几乎同样高度，设计得精巧而又实用。将"天把式"架起后，梨农们便可在梨园中自如料理梨树。由此，什川果农形成了独特的栽培方式与农耕文化。

什川农民固守着传统的农业生产经验和技术，即使在科技发达的今天，"天把式"依然是什川梨园重要的农业生产工具，帮助什川农民使梨园历经数百年后仍保持着一定的经济产量。

地理位置：北纬36°05′—36°50′、东经103°32′—104°22′，地处甘肃省兰州市皋兰县。

气候特点：温带半干旱气候。

认定时间：2013年被农业部列为首批中国重要农业文化遗产。

Geographical location: 36°05′-36°50′N, 103°32′-104°22′E; located in Gaolan County, Lanzhou City, Gansu Province.
Climate type: temperate semi-arid climate.
Time of identification: In 2013, it was listed as one of the first batch of the China-NIAHS by the Ministry of Agriculture.

Pear Tree Cultivation System with the most Stock of Ancient Pear Trees
Gaolan Shichuan Ancient Pear Orchard

The Shichuan Ancient Pear Orchard, known as the "Best Ancient Pear Orchard in the World", is located in Gaolan County, Gansu Province. Located in the Loess Plateau, Gaolan County is surrounded with mountains and gullies. The dry climate brings long illumination time while little rainfall: the average annual evaporation is almost six times that of precipitation. Geographically, however, Shichuan is an "enclave" of Gaolan County.

The time-honored Shichuan, endowed by nature with men of talent vegetation of rarity, is a masterpiece of the Yellow River culture. Shichuan is about 20 kilometers east of Lanzhou City and 20 kilometers west of Gaolan County. It has a total area of 405 square kilometers and an average elevation of 1,500 meters. The moist climate nourishes the beautiful scenery and makes Shihuan get the reputation of the "Township of Fruits and Vegetables".

Shichuan Township is built on the banks of Yellow River and taken by the Yellow River in her arms like a child. Shichuan is surrounded by water in three sides and embosomed in hills. It enjoys mild climate and fertile soil. Shichuan introduced pear trees and cultivated them as cash fruit trees during the years reigned by Emperor Jiajing period of the Ming Dynasty. In the relatively arid environment, Shichuan people adapted to the local conditions and built waterwheel-like tool to diverse the Yellow River water for pear trees irrigation. The pear trees that has demanding requirement for water have survived in Shichuan. The unique geographical environment provides a suitable growth space for pear trees, and pear trees grow strong here. Since then, pears have become the main cash crop of Shichuan. Shichuan ancient pear orchard is a composite agro-ecosystem featuring thousands of pear trees of more than 200 years old. The total area of the core area of the heritage site is about 42 square kilometers.

Not only the geographical environment that is not easily disturbed by the outside and such a large-area, but also the cultivation method and the management of pear orchard that have been passed down from generation to generation contiguous make the ancient pear orchard preserved well until this day. Fruit farmers not only need to loosen and fertilize the soil, "scrape the bark" in early spring, " stack sand" for pest control in the flowering period, but also need to use a "sky-scraping" ladder to prune trees, thin the flowers and fruits, tidy branches, and pick fruits, which forms a unique cultivation method and farming culture. Shichuan people call peat trees "high fields" and the picking tool a "sky-scraping" ladder because growers have to cultivate "in the air". The ladder is almost the same height as the crown, and the design is exquisite and practical. The pear farmers can manage pear trees easily with the "sky-scraping" ladder.

Shichuan farmers adhere to the traditional agricultural production experience and technology. Even with the rapid development of science and technology, the "sky-scraping" ladder is still an important agricultural production tool for Shichuan people. Hundreds of years later, the Shichuan pear orchard still maintains a high yield.

什川古梨园位于甘肃省兰州市近郊，处在黄河之滨，这里现存百年以上的古梨树 2000 多株。2013 年被正式录入"世界吉尼斯纪录大全"。（任世琛 摄）

Shichuan Ancient Pear Orchard is located in the suburbs of Lanzhou City, Gansu Province, on the bank of the Yellow River. There are more than 2,000 ancient pear trees that have been in existence for more than 100 years. In 2013, it was officially listed into the Guinness World Records. Photographed by Ren Shichen

什川古梨树栽培历史悠久，自明嘉靖年间（1522—1566年），当地果农仿建水车汲黄河水灌溉田园，开始栽植梨树。现存古梨树的树龄大多在200年以上，至今仍然硕果累累。图为什川美景和深秋的古梨园。（上图 任世琛 摄；下图 武翠敏 摄）

Shichuan ancient pear trees cultivation has a long history. Since the years reigned by Emperor Jiajing in Ming Dynasty, the local fruit farmers built a diversion tool with the waterwheel as its original to irrigate the field and began to plant pear trees. Most of the existing ancient pear trees have been in existence for more than 200 years, and they are still fruitful. The pictures show the beautiful scenery of Shichuan and the ancient pear orchard in late autumn. The picture above photographed by Ren Shichen, and below by Wu Cuimin

什川梨以软儿梨最为有名，是严冬季节深受人们喜爱的梨中珍品。软儿梨清香、醇甜，尝一口顿感神清气爽，舒畅不已。梨树树干高大，而栽培梨树需要修枝整形、疏花疏果、采摘果实，因此什川人将栽培梨树叫作"种高田"。"种高田"所需要的工具云梯被称作"天把式"。图为什川软儿梨和当地果农用"天把式""种高田"。（上图 张志奎 摄；下图 张耘 摄）

Among all the pear varieties of Shichuan, soft pear is the most famous one. It is deeply affected by people in winter. The soft pear is sweet and full of fragrance. The fresh and cool taste must refresh you completely. Shichuan people call peat trees "high fields" because growers have to cultivate "in the air". The tool required for "high fields" is called a "sky-scraping" ladder. The pictures show the soft pear and the local fruit farmers who are using the "sky-scraping" ladder "cultivate" the "high fields". The picture above photographed by Zhang Zhikui, and below by Zhang Yun

什川古梨园，四季景色美不胜收。春夏之交，梨园梨花似雪，翠盖参天，生机盎然，置身梨园如入"天然氧吧"，令人心旷神怡；经金秋梨压枝头，进入严冬的古梨园，古树虬枝，加上白雪、红叶，百年古梨园在什川人的照料下，持续焕发着生机。（左图 任世琛 摄；右图 武翠敏 摄）

Shichuan Ancient Pear Orchard teems with stunning sceneries in all seasons. At the turn of spring and summer, pear flowers are as white as snow and the crowns are high enough to touch the sky. People who ever entered there will feel like into a "natural oxygen bar" and be refreshed. In golden autumn, the branches will be crowded with plump pears; in winter, the ancient trees, as well as the white snow, red leaves, are all under good care of Shichuan people and full of vitality. The picture left photographed by Ren Shichen, and right by Wu Cuiming

什川古梨园是当地重要的旅游资源,当地政府依托古梨树资源,已连续举办了十多届"兰州·什川之春"旅游节,把旅游观光、文体娱乐等融为一体,形成以梨园美景观赏、黄河风光游览、农家休闲娱乐等为主的新型休闲农业旅游区。
任世琛 摄

Shichuan Ancient Pear Orchard is an important tourism resource in this area. The local government has successively held the tourism festival named "Lanzhou·Shichuan Spring" more than ten times based on the resources of ancient pear trees, which integrates tourism, entertainment and entertainment into one and forms a new type of leisure agriculture with the pear orchard, the Yellow River and the agritainment as its cores. Photographed by Ren Shichen

农、林、牧循环复合生产体系
甘肃迭部扎尕那农林牧复合系统

扎尕那位于陇、青、川三省交界之处的甘肃省迭部县西北，包含四村一寺。

在藏语中，"扎尕那"的意思是"石匣子"。这是对藏族村寨扎尕那的地理位置特征最形象的比喻——环绕三面的群山，为扎尕那塑造了天然的、坚不可摧的"城墙"。扎尕那汇集了山川、峡谷、森林、冰川遗迹等多种地貌，是高寒草原、温带草原和暖温带落叶林三种植被类型的过渡地带。独特的地理位置、丰富的植被类型，给扎尕那的农业带来了更多的可能。扎尕那的人们将林木与人工栽培的经济作物以及动物按照一定的时间和空间，有机地排列在一起，使得农业、林业、畜牧业混合发展，在获得经济效益的前提下，还能保护当地生态环境不受破坏。

对扎尕那农林牧复合系统追根溯源，可以回溯到3000年前。当时，迭部县已经出现了畜牧文明的萌芽。3世纪初，蜀国伐魏，迭部归蜀国管辖，蜀将姜维在沓中（今迭部扎尕那境内）休整兵力、积蓄粮草，遂把内地的农业文明引入到此，扎尕那开始了畜牧业与农耕业并存的态势。到了吐谷浑统治时期（4—7世纪，吐谷浑是鲜卑慕容一支，东晋十六国时期控制了青海、甘肃等地，与南北朝各国都有友好关系），扎尕那的农林牧复合系统开始崭露头角。明代中期（16世纪初）以后，社会稳定，扎尕那的治理者主动向内地学习先进的农业文化，这里的农业得以和畜牧业齐头并进，逐渐形成了农林牧复合发展的农业系统。这种复合发展体现在三个方面：农牧复合、农林复合及林牧复合。扎尕那的农民在天然草地上开垦农田，使得耕地与天然草地相间分布，农业耕种与畜牧养殖混合存在，家畜是耕种的重要劳动力，农作物的废料又可以为家畜提供饲料；在耕地之间，农民种上少量林木，可以为日常生活生产提供材料，而外围的原始森林，就得以保留下来，起涵养水源等生态作用；此外，当地农民还在林下进行畜牧养殖，家畜的粪便是树木生长最好的养分。

在扎尕那的农林牧复合系统内，含有多个生物种群，产出多种产品，并可以获得高效收益。这种方式，在高寒地区对于维护生态平衡和保障生态安全有着重要的意义。目前，扎尕那的植被覆盖率已达到87%，是该地区重要的水源涵养区，对维护高寒地区的生态环境和生物多样性发挥着巨大的作用。

地理位置：北纬33°41′20″—34°17′30″、东经103°00′37″—104°04′35″，地处甘肃省甘南藏族自治州迭部县。
气候特点：非典型性大陆性气候。
认定时间：2013年被农业部列为首批中国重要农业文化遗产，2017年被联合国粮农组织列为全球重要农业文化遗产。

Geographical Location: 33°41′20″-34°17′30″N, 103°00′37″-104°04′35″E; located in Diebu County, Gannan Tibetan Autonomous Prefecture, Gansu Province.
Climate type: atypical continental climate.
Time of identification: In 2013, it was listed as one of the first batch of the China-NIAHS by the Ministry of Agriculture. In 2017, it was listed as one of the GIAHS by the FAO.

Cyclic Agriculture-Forestry-Animal Husbandry Composite System
Diebu Zagana Agriculture-Forestry-Animal Husbandry Composite System

Zagana is located in the northwest of Diebu County, Gansu Province, at the junction of the three provinces of Gansu, Qinghai and Sichuan. It consists of four villages and one temple.

Zagana means "stone casket" in Tibetan. This is the most vivid description for the geographical location of the Zagana Village. The mountains surrounding in three sides form a natural and indestructible "wall" for it. There are a variety of landforms such as mountains, canyons, forests, and glacial relics. It is a transitional zone between three types of vegetation: alpine grassland, temperate grassland and warm temperate deciduous forest. The unique geographical location and rich vegetation types bring more possibilities to the agriculture of Zagana. People in Zagana organically arrange forests, cultivated cash crops and animals according to a certain time and space, so that agriculture, forestry and animal husbandry can be developed circularly, which protects the ecological environment under the premise of obtaining economic benefits.

Zagana Agriculture-Forestry-Animal Husbandry Composite System can be traced back to 3,000 years ago. At that time, the sprout of animal husbandry had already appeared in Diebu County. At the beginning of the 3rd century, the Shu Kingdom sent a punitive expedition against the Wei Kingdom and captured Diebu County. The military general Jiang Wei and his soldiers rested and accumulated military supplies in Diebu County, and then introduced the agricultural civilization of the mainland to this place. Animal husbandry and agriculture were in coexistence since then. In the period of Tuyuhun's reign (the 4th and 7th centuries, Tuyuhun was one branch of Murong Family. It reigned over Qinghai and Gansu during the Sixteen Kingdoms period in the East Jin Dynasty, and had friendly relations with the countries of the Northern and Southern Dynasties), the agriculture-forestry-animal husbandry composite system began to emerge. After the middle period of the Ming Dynasty (early 16th century), the society was stable, and the governors of Zagana took the initiative to learn advanced agricultural practice from the mainland. The agriculture here and the animal husbandry industry went hand in hand, and gradually formed a comprehensive system of agriculture, forestry and animal husbandry. This compound development is reflected in three aspects: the compound of agriculture and animal husbandry, agriculture and forestry, and forestry and animal husbandry. The farmers in Zagana reclaimed farmland on natural grassland, so that the cultivated land and natural grassland were distributed alternatively. Agricultural farming and animal husbandry were an organic whole. the livestock was an important labor force for farming, and crop wastes could provide feed for the livestock. Between cultivated lands, farmers plant a small amount of forest trees, which could provide materials for daily life production, while the peripheral virgin forests would be preserved and cultivated to conserve water resource. In addition, local farmers carried out animal husbandry under the forests, and the livestock manure is the best nutrients for tree growth.

Zagana Agriculture-Forestry-Animal Husbandry Composite System contains multiple biological populations and can produce a variety of products. It is of great significance for maintaining ecological balance and ensuring ecological security. At present, the vegetation coverage rate in Zagana has reached 87%. It is an important water conservation area in the region and plays an indispensable role in maintaining the ecological environment and biodiversity in alpine regions.

扎尕那农林牧复合系统核心区位于甘肃省甘南藏族自治州迭部县益哇乡。在该系统中，农、林、牧之间的循环复合使其生产能力和生态功能得以充分发挥，游牧、农耕、狩猎和樵采等多种生产活动的合理搭配使劳动力资源得到充分利用。汉地农耕文化与藏区游牧文化的相互交融形成了特殊的农业文化。图为扎尕那山寨。（郭群 摄）

The core area of Zagana Agriculture-Forestry-Animal Husbandry Composite System is located in Yiwa Township, Gannan Tibetan Autonomous Prefecture, Gansu Province. In this system, the circulation of agriculture, forestry, and animal husbandry allows the production capacity and ecological functions to be fully utilized. The reasonable combination of various production activities such as nomadism, farming, hunting, and firewood cutting makes full use of labor resources. The fusion of farming culture of Han and Tibetan nomadism has formed a special agricultural culture. The picture shows Zagana Village. Photographed by Guo Qun

农林牧复合系统是指在同一土地管理单元上，人为地把多年生木本植物（如乔木、灌木、棕榈、竹类等）与其他栽培植物（如农作物、药用植物、经济植物以及真菌）及动物在空间上或按一定的时序有机地排列在一起，形成具有多种群、多层次、多产品、多效益特点的人工生态系统。在干旱缺水地区，农、林、牧复合系统可发挥其生态优势，林木系统的林冠可以截留降水，枯枝落叶层及地被层可使降水渗入土层，减少表面径流和土壤冲刷，增加土壤湿度。（闵庆文 摄）

The composite system refers to artificially organically manage perennial woody plants (such as arbors, shrubs, palms, bamboos, etc.), cultivated plants (such as crops, medicinal plants, cash plants and fungi) and animals on the same land management unit based on the space or certain time sequence to form an artificial ecosystem with multiple groups, levels, products and benefits. In water-deficient areas, the agriculture-forestry-animal husbandry composite system can exert its ecological advantages. The crowns of the forest system can intercept the rainfalls, and the withered leaves and the ground can make the rainfalls seep into the soil layer, which can help reduce the overland runoff and soil erosion, and increase the soil moisture. Photographed by Min Qingwen

西北地区·Northwest China/甘肃省·Gansu Province

扎尕那农林牧复合系统位于高寒草原、温带草原和暖温带落叶林三大植被气候类型的交汇处，独特的地理区位为农林牧复合经营提供了自然资源和经济社会基础。农田、河流、民居、寺庙与周围的山林和草地互相映衬，滩地耕种、林草相间，呈现出农、林、牧相互依存、优势互补的复合生产方式。上图为当地农民在田间耕作，下图为当地具有民族特色的水力转经筒。（郭群 摄）

The composite system is located at the intersection of the three major vegetation types of alpine grassland, temperate grassland and warm temperate deciduous forest. The unique geographical location provides natural resources and economic and social foundation for the combination of agriculture, forestry and animal husbandry. Farmland, rivers, houses, and temples harmonize with the surrounding forests and grasslands, showing a composite production mode in which agriculture, forestry, and animal husbandry are interdependent and complement one another. The picture above shows the local farmers farming in the field and the below shows the local ethnic hydropower water-diverting cylinders. Photographer by Guo Qun

扎尕那居民以藏族为主，村子里的男人大部分会说普通话，女人大部分不会说普通话，村里最主要的出行方式是骑马和乘拖拉机。图为当地背经筒的藏族妇女以及在编制牛饰的藏族妇女。（郭群 摄）

The residents of Zagana are mainly Tibetans. Most of the men in the village could speak Mandarin while most of the women do not understand it. The main trip mode in the village is horses and tractors. The pictures show Tibetan women backing cylinders and Tibetan women who are weaving cattle-shaped ornaments. Photographed by Guo Qun

这里地处高寒贫瘠的生态脆弱地区，又是生物多样性保护优先区域，还是长江与黄河分水岭的上游地带，是重要的水源涵养区，对维护生态平衡和保障生态安全具有重要作用。可以说，独特的生态区位促进了这里游牧文化、农耕文化与藏传佛教文化的融合与发展，造就了独特的扎尕那农、林、牧复合系统。图为扎尕那民居景观。（闵庆文 摄）

The system is located in a cold and barren ecologically fragile area where it is a priority area for biodiversity conservation, also an upstream area of the Yangtze River and the Yellow River watershed, an important water source conservation area and plays a significant role in maintaining ecological balance and ensuring ecological security. The unique ecological location promotes the integration and development of nomadic culture, farming culture and Tibetan Buddhism culture, and creates the unique agriculture-forestry-animal husbandry composite system. The picture shows the scene of houses in Zagana. Photographed by Min Qingwen

"千年药乡"的杰出代表
甘肃岷县当归种植系统

当归当归，"正是归时又不归"。除了是一味药材外，当归还被寄予了更多的含义。李时珍在《本草纲目》中如此解释"当归"的命名：古人娶妻要嗣续也，当归调血为女人要药，有思夫之意，故有"当归"之名。如此看来，"当归"的命名应当与它的药效有关。

当归是伞形科多年生草本植物。根肥大，叶为羽状复叶，夏季开白色花，复伞形花序，果实长椭圆形，侧棱有广翅。中医学上以根入药。其主根呈圆柱形或圆锥形，尾端渐细；表皮棕褐色或黄褐色，主根上端有不甚明显的环形皱纹；横断面是白色或淡黄棕色，有线状纹理，习称"菊花心"，中心有的有白色髓心；体质坚硬，吸潮后软韧。

当归又被称为"草头归"，其根茎在中医中的广泛应用，常常使人们忘记了这是一种可以长到1米高的草本植物。当归入药已有上千年的历史。中医认为，当归有调气养血、调经止痛的作用，是一味治疗妇科病的重要药材。现存最早的中药学著作《神农本草经》中，便有对当归治疗妇女疾病方法的记载。

当归喜欢凉爽的环境，我国的甘肃、云南、四川等高海拔地区均有栽培。其中，甘肃岷县是当归的主产区，被称为"当归之乡"。甘肃岷县当归种植系统即以当归栽培与利用为特色的复合农业文化系统。

岷县位于黄河上游，是甘南草原、黄土高原、陇南山地的交汇地带，平均海拔超过2000米。这里气候凉爽、降水量丰富、光照时间长，对于喜好低温和光照的当归来说，是生长的好地方。岷县有着1700多年的当归种植栽培历史，人们习惯称呼这里生产的当归为"岷归"。成书于汉末的《名医别录》中，就有对甘肃岷县地区种植当归的记载，这几乎是关于当归产地最早的记载。

岷县生产的当归，有特异的芳香味道，微甘而稍苦，肉多汁少，气味香醇，含有挥发性和水溶性物质106种，被国外誉为中国的"妇科人参"。

在上千年围绕着当归种植栽培、生产炮制的日子中，岷县人的节庆习俗、商贸、饮食、建筑、服饰、耕作习惯早已与当归息息相关，形成了岷县独特的当归农耕文化。

地理位置：北纬34°07′34″—34°45′45″、东经103°41′29″—104°59′23″，地处甘肃省定西市岷县。
气候特点：温带半湿润向高寒湿润气候过渡带。
认定时间：2014年被农业部列为第二批中国重要农业文化遗产。

Geographical location: 34°07′34″-34°45′45″N, 103°41′29″-104°59′23″E; located in Min County, Dingxi City, Gansu Province.
Climate type: transitional zone from temperate semi-humid to alpine humid climate.
Time of identification: In 2014, it was listed as one of the second batch of the China-NIAHS by the Ministry of Agriculture.

An Outstanding Representative of Millennium Medicine Township
Minxian Angelica Planting System

Angelica has the same pronunciation of Chinese characters "Dang Gui(当归)", which means that it is time to return your home. In addition to be a medicinal herb, angelica has been given more meanings. Li Shizhen explained the naming of angelica in the *Compendium of Materia Medica* like this: the housewives took an important mission in one's family, which was giving birth to an heir. The angelica was a useful medicine for them to enrich the blood, and there was also a meaning of an expectation of the return of their husbands. It seems that the naming of angelica should be related to its efficacy.

Angelica is a perennial herb of the umbelliferae family. The roots are hypertrophy, and the leaves are pinnately compound leaves. In summer, white flowers bloom in compound corymb inflorescence; the fruit is long and oval, and the side edges have broad-winged. Roots are the part that can be utilized as medicine. The main root is cylindrical or conical, and the tail end is tapered; the epidermis is tan or yellowish brown, and the upper end of the main root has less obvious annular wrinkles; the cross section is white or yellowish brown and wrinkled and has white pith which is usually called "chrysanthemum heart"; the main root is hard, while turns to be pliable after moisture absorption.

Angelica is also known as "grass angelica", and its roots are widely used in traditional Chinese medicine. It is often forgotten by us that it is a kind of herb that can grow up to 1 meter high. It has been in use as Chinese medicine for a thousand years. Doctors of Chinese medicine believes that angelica has the effect of regulating vital energy and nourishing blood, regulating menstruation and relieving pain, so it is an important medicine for treating gynecological diseases. There are records of the treatment of gynecological diseases in the earliest existing Chinese pharmacy book *Shen Nong's Herbal Classic*.

Angelica likes the cool environment and is planted in high-altitude areas like Gansu, Yunnan, Sichuan and other regions. Among them, Min County in Gansu Province is the main producing area, which is called "the Hometown of Angelica". The Gansu angelica plantation system is a compound agricultural system featuring cultivation and utilization of angelica.

Min County is located in the upper reaches of the Yellow River. It is the intersection of Gannan Grassland, Loess Plateau and Longnan Mountain. The average elevation is over 2000 meters. The climate here is cool, the precipitation is abundant, and the lighting time is long. For Angelica, which prefers low temperature and light, it is a good place to grow.

Min County has a history of cultivation of more than 1,700 years. People are used to calling the angelica produced in Min County as "the Angelica of Min". There is record of angelica planting in Min County in the book *Supplementary Records of Famous Physicians* written in the late Han Dynasty, which is almost the earliest record of the origin of angelica.

Angelica produced in Min County has a specific aromatic flavor. It is slightly sweet and slightly bitter, less juicy while meaty and fragrant. It contains 106 kinds of volatile and water-soluble substances, and is praised by foreigners as the "gynecological ginseng of China".

In the days of planting, cultivation and production of angelica in millennium, the festival customs, trade, food, architecture, costumes, and farming habits of Min people have long been closely related to angelica, forming the unique angelic farming culture of Min County.

甘肃岷县当归种植系统位于甘肃省定西市西南部，正处于陇中黄土高原、甘南草原和陇南山地接壤区，位居定西、甘南、陇南、天水四区（州、市）几何中心，享有陇原"旱码头"美称，是"茶马古道"重镇和甘肃南部重要的商品集散地。图为岷县当归育苗基地。（刘毅 摄）

The Minxian Angelica Planting System is located in the southwest of Dingxi City, Gansu Province, the border area of the Loess Plateau, Gannan Grassland and Longnan Mountain, and the geometric center of Dingxi, Gannan, Longnan and Tianshui City. It is known as the "dry pier", an important commodity distribution center in southern Gansu once passed through by the "Ancient Tea Horse Road". The picture shows the angelica seedling base in Min County. Photographed by Liu Yi

岷县当归产自海拔 2000 米以上的高寒阴湿山区,早在 1700 多年前,该县就有了种植当归的记载,当地群众在长期的生产实践中积累了极为丰富的当归种植经验。图为当地农民在挖当归。(石志平 摄)

Angelica of Min County is produced in the cold and humid mountainous area more than 2000 meters above sea level. As early as more than 1,700 years ago, the county began to record the cultivation of angelica. The local people have accumulated a wealth of experience in the cultivation of angelica in the long-term production practice. The picture shows local farmers digging for Angelica. Photographed by Shi Zhiping

"岷归"气有特异芳香,味微甘而稍有苦辛,肉多汁少,气味香醇。其因产量大、质量佳、销量广而驰名中外,不仅享誉中国的港、澳、台地区,还远销东南亚、欧美等 20 多个国家和地区,被誉为"妇科人参"。经过多年发展,岷县已形成了当归种植、质量安全、生产技术、产地环境等一整套种植的规范流程和标准,其无公害标准化示范基地达 10 多万亩。图为岷归的晾晒。(王喜生 摄)

"The Angelica of Min" has a specific aroma. It is slightly sweet, slightly bitter, less juicy while meaty and fragrant. Because of its high field, good quality, huge market and brand awareness, it is well-known as "gynecological ginseng" and exported to more than 20 countries and regions such as Hong Kong, Macao and Taiwan, Southeast Asia, and European and American countries. After years of development, Min County has formed a set of standardized procedures of plantation and techniques and requirements of quality, safety, production environment. Its pollution-free standardization demonstration base has reached more than 100,000 mu. The picture shows the drying of the Angelica of Min. Photographed by Wang Xisheng

千百年来，当归种植在岷县这块土地上被传承发展，成为岷县农民的主要经济来源和富民强县的支柱产业。独特的栽培条件、传统的栽培技术以及加工炮制技术造就了"岷归"品牌，"岷归"已成为我国中医药文化遗产的重要组成部分。图为当归被削制加工成药片状和家庭熏当归。（石志平 摄）

Angelica has been in development on this land for thousands of years, it becomes the main source of income for angelica growers and the pillar industry of Min County. The unique environment, traditional cultivation and processing techniques have created the brand of "the Angelica of Min", which is an important part of the cultural heritage of Chinese medicine now. The pictures show the processing of angelica. It is cut into strips and home-smoked. Photographed by Shi Zhiping

特殊的区位造就了独特的自然人文景观、农耕文化和民俗文化。岷县人的习俗、节庆、商贸、饮食、建筑、服饰、耕作习惯等无不与当归息息相关。图为当地二郎山花儿会景象。（陈旭萍 摄）

The special location creates a unique natural human landscape, farming culture and folk culture. The customs, festivals, commerce, food, architecture, clothing, farming habits, etc. of Min people are all closely related to angelica. The picture shows the Flower Temple Fair held in Erlang Mountain. Photographed by Chen Xuping

"二月二"春台会是岷县城乡人民共同欢度的盛大节日,一年一度的"二月二"祭祀庙会热闹非凡、极具魅力,前去赶会献蜡的游人川流不息,人们穿着过年的节日盛装,全家老小出门浪会观光,俗称"浪二月二"。图为当地的二月二庙会现场。(陈旭萍 摄)

The "Lunar February 2nd Festival" is a grand festival shared by all the urban and rural people of Min County. The annual temple fair attracts the tourists streaming in and out. No matter the elders and the children will dress up to attend the grand festival. The picture shows the temple fair. Photographed by Chen Xuping

独特环境、独特品种、独特技艺

宁夏灵武长枣种植系统

宁夏自古出枣。明代弘治年间（1488-1505年），枣树开始在宁夏灵武、中宁、中卫等地广为栽培。独特的地理环境和自然气候条件，使宁夏出产的枣个儿大、肉厚、脆甜可口。其中，属灵武长枣的品质为最佳。

灵武，古称灵州，宁夏回族自治区银川市下辖县级市，是宁夏回族自治区和银川市工业发展的核心区域。这里地处宁夏中部的黄河东岸，海拔1250米。典型的大陆性气候，使灵武四季分明，光热充足，年均降水量200毫米而又蒸发强烈。但是，灵武素来是出产水果的地方。郦道元在《水经注》中这样描述灵州（灵武古地名）："桑果余林，仍列州上。"诗人韦蟾称赞灵州是"水果之乡"，并写下"贺兰山下果园成，塞北江南旧有名"的诗句。

这是因为与其他内陆城市相比，灵武有着明显的农业生产优势：黄河在旁，灌溉水源丰富。这一优势条件，使得灵武有着"塞上江南、水果之乡"的美誉。

灵武的自然环境条件对于枣树的生长来说是再好不过了：坐果时昼夜温差大而又降水少，可以更多地保留果实体内的糖分；沙壤的土质，能够给枣树提供充足的营养成分；黄河水可以保证枣树不被干旱的气候所累。

据资料显示，灵武长枣有着超过1300年的栽培历史。唐代时，灵武长枣因品质上乘曾作为贡果向皇室进贡，是"果中珍品"。宁夏灵武长枣种植系统即以皇家贡品长枣为特色的农业文化系统。

喝着黄河水长大的灵武长枣，抗逆性强，果实营养丰富，药用价值高。长枣呈椭圆形，最大的枣重量一颗可达40克，口感酸甜，汁甜味美。作为灵武的独有品种，长枣现在在灵武的种植面积达到14.2万亩，让灵武获得了"中国长枣之乡"和"全国枣产业十强县"的称号，支撑了灵武市的农业产业。

地理位置：北纬37°60′—38°01′、东经105°59′—106°37′，地处宁夏回族自治区银川市灵武市。
气候特点：温带大陆性季风气候。
认定时间：2014年被农业部列为第二批中国重要农业文化遗产。

Geographical location: 37°60′-38°01′ N, 105°59′-106°37′ E; located in Lingwu City, Yinchuan City, Ningxia Hui Autonomous Region.
Climate type: temperate continental monsoon climate.
Time of identification: In 2014, it was listed as one of the second batch of the China-NIAHS by the Ministry of Agriculture.

Unique Environment, Variety and Artistry
Lingwu Long Jujube Planting System

Jujubes have been produced in Ningxia since ancient times. During the Hongzhi period of the Ming Dynasty, jujube trees began to be widely cultivated in places such as Lingwu, Zhongning and Zhongwei of Ningxia. The unique geographical environment and natural climate conditions make the jujubes produced in Ningxia big, fleshy, crisp and sweet. Among them, Lingwu Long Jujube enjoys the best quality.

Lingwu, also known as lingzhou in ancient times, is a county-level city under Yinchuan City, capital of Ningxia Hui Autonomous Region. It is the core area of industrial development of Ningxia Hui Autonomous Region and Yinchuan City. Located on the east bank of the Yellow River in central Ningxia, it is 1250 meters above sea level. The typical continental climate makes Lingwu distinct in four seasons, abundant in light and heat, with an average annual precipitation of 200 mm and strong evaporation. However, Lingwu has always been the place which is celebrated for fruits. Li Daoyuan, an oppressive official and geographer in the Northern Wei Dynasty of China, described Lingzhou (as Lingwu then was called) in his *Commentary on the Waterways Classic*: "Mulberry trees and many other fruit trees are still standing here until today". Poet Wei Chan praised Lingzhou as "the Land of Fruits", and wrote down the verses: "The orchard at the foot of Helan Mountain being completed, the Southern Frontier for long is well-noted".

This is because compared with other inland cities, Lingwu has obvious advantages in agricultural production: beside the Yellow River, there are abundant irrigation water sources. This advantage makes Lingwu known as "the Land of Fruits and the Southern Frontier".

Lingwu's natural environment conditions are unequalled for the growth of jujube trees: when setting fruits, the large difference in temperature between day and night and low amount of precipitation help retain more sugar in the fruit; sandy soil can provide adequate nutrients for jujube trees; and the water of the Yellow River can ensure that jujube trees are not affected by the dry climate.

According to the data, the cultivation of Lingwu Long Jujube has enjoyed a history of more than 1300 years. In the Tang Dynasty, Lingwu Long Jujube was once used to pay tributes to the royal family because of its superior quality. It was a "Treasure in Fruits". Lingwu Long Jujube Planting System of Ningxia is an agricultural and cultural system which is characterized by royal tribute long jujube.

Being irrigated with the water of the Yellow River, Lingwu Long Jujube has strong resistance, rich nutrition and high medicinal value. Being oval in shape, long jujube, the largest of which weighs up to 40 grams, tastes sour and sweet, juicy and delicious. As a unique variety of Lingwu, the planting area of long jujube in Lingwu now reaches 142,000 mu, which has earned Lingwu the title of "Land of Chinese Long Jujube" and "China's Top Ten County of Jujube Industry ", thus providing strong support for the agricultural industry of Lingwu City.

灵武长枣从唐朝开始就被列为皇室贡品，距今已有1300多年的栽培历史。2003年以来，灵武长枣这一古老的优良品种得到大规模发展，种植面积达到14.2万亩。图为宁夏灵武市世界枣树博览园航拍图。（季正 摄）

Lingwu Long Jujube has been listed as a royal tribute since the Tang Dynasty. It has been cultivated for more than 1300 years. Since 2003, Lingwu Long Jujube, an ancient fine variety, has been developed on a large scale, with the planting area reaching up to 142,000 mu. The picture shows the aerial photograph of the World Jujube Expo Park in Lingwu City, Ningxia. Photographed by Ji Zheng

灵武长枣是长期自然筛选出来的具有地方特色的鲜食珍品，抗逆性强，果实营养丰富，药用价值高，发展潜力大。目前，灵武长枣产品市场广阔，形成了较好的品牌效应。（宋克强 摄）

Lingwu Long Jujube is a kind of fresh food treasure with local characteristics, which has been naturally screened for a long time. It has strong resistance, rich nutrition, high medicinal value and great potential for development. At present, Lingwu Long Jujube products have a broad market, forming a good brand effect. Photographed by Song Keqiang

灵武长枣性状优良，果实成熟期在9月下旬至10月上旬，果个大，长椭圆形或圆柱形，大小较整齐。灵武长枣鲜食果味鲜美，质地酥脆，汁液多，果肉呈白绿色，属功能食品，营养价值高。图为第四届灵武长枣节赛果会上等待被评价的灵武长枣。（宋克强 摄）

Lingwu Long Jujube has excellent commercial properties. The ripening period is from late September to early October. Being oblong or cylindrical in shape, the fruit is large in size and has relatively regular size. With its white- green pulp, the fresh Lingwu long jujube, being delicious, crisp, and juicy, is a functional food with high nutritional value. The picture shows Lingwu Long Jujube waiting to be evaluated at the 4th Lingwu Long Jujube Festival. Photographed by Song Keqiang

为发展灵武长枣，打造地方品牌，灵武市通过扩大种植规模，优化品种结构，实施无公害标准化生产，培育、壮大长枣储运和深加工龙头企业，加快灵武长枣良种选育，加大基地标准化技术推广力度，建立完善的营销体系等方面来推动灵武长枣产业发展。图为灵武长枣的分拣装箱、遴选加工和评比。（宋克强 摄）

In order to develop Lingwu Long Jujube production, build local brands and push the industrial development, Lingwu City cultivates and expands leading enterprises of storage, transportation and deep processing of long jujube by expanding planting scale, optimizing variety structure, implementing pollution-free standardized production, accelerates the breeding of Lingwu Long Jujube, strengthen the promotion of base standardization technology and establish a sound marketing system. The pictures show the sorting, packing, selection, processing and evaluation of Lingwu Long Jujube. Photographed by Song Keqiang

为了传承和保护好长枣这一农业文化遗产，目前灵武市制定了系列保护办法，建成了"世界枣树博览园"，成立了灵武市世界枣树博览园管理中心，并定期举办长枣文化节。图为第五届灵武长枣文化节开幕式盛况。（宋克强 摄）

In order to inherit and protect this agri-cultural heritage, Lingwu City has formulated a series of protection measures, built the World Jujube Expo Park, established the Lingwu World Jujube Expo Park Management Center, and regularly held Long Jujube Cultural Festival. The picture shows the opening of the 5th Lingwu Long Jujube Cultural Festival. Photographed by Song Keqiang

大型地下农业水利灌溉工程
新疆吐鲁番坎儿井农业系统

在新疆的吐鲁番地区，坎儿井可以被称为"生命井"。

吐鲁番深居中国大陆内部，温带荒漠气候让这里炎热而又少雨。与年均3000毫米以上的蒸发量相比，16毫米的降水量几乎可以被忽略不计。极端干燥的气候，让吐鲁番有着"火洲"之称。这样的环境看起来并不适宜人类生存和农业发展，但是，吐鲁番偏偏还是沙漠中的绿洲。吐鲁番四面环山，西有阿拉山、北有博格达山脉、南邻库鲁塔格山，四周高、中间低，是典型的盆地地形。每年夏季，气温升高以后，四周山脉的积雪便开始消融，产生大量的融水。冰雪融水渗透地下，给吐鲁番带来丰富的地下水。

人们对大自然的馈赠总是有着许多使用的智慧。吐鲁番地区的人们沿着地面坡度，在地下开挖水渠，并最终将地下水引到农田和民居，这便是坎儿井，是世界上最大的地下水利灌溉系统之一。新疆吐鲁番坎儿井农业系统就是在干旱荒漠地区运用独特方式、引高山雪水入地下潜流、灌溉农田的农业水利生产系统。

坎儿井是利用地面坡度，引用地下水到农田和民居的一种独具特色的地下水利工程，由竖井、暗渠、明渠和涝坝四部分组成。人们在地面上打造竖井，井底与暗渠相通，井深随着地面坡度的下降而递减。除集水和串联暗渠功能外，竖井还有着通风出土、供施工和维修人员上下的功能；暗渠在发挥输水功能外，也可以起到集水的作用；明渠则是根据坡度的下降，在低地开挖的直接引水区；涝坝是坎儿井的终点，地下水由暗渠流进明渠后，最终进入涝坝中得以保存。整套流程，完全依靠水的自流实现引水。把水贮存在地下，也减少了当地蒸发量大带来的水量损耗。其优良的水质可供农田灌溉和人畜饮用。

坎儿井在中国最早见于汉代，《史记·河渠书》等多部汉代书籍对此均有记载。当时，今新疆、甘肃、陕西一带已经开始利用这种方式开采地下水。不过，具体到吐鲁番，坎儿井来源于何地、何时开挖，至今众说纷纭。但是，坎儿井最终在吐鲁番得以延续、利用和保存下来。这种地下水利灌溉系统在全球多个干旱地区可见。最多时，吐鲁番拥有超过1000条坎儿井，年流量5.6亿立方米，灌溉面积约35万亩，在极其严酷的气候环境条件下，给吐鲁番的农业带来生机，养活了吐鲁番地区一代又一代的人，让吐鲁番这个"火洲"成为沙漠上的"绿洲"。

地理位置：北纬41°12′—43°40′、东经87°16′—91°55′，地处新疆维吾尔自治区吐鲁番市。

气候特点：温带大陆性干旱荒漠气候。

认定时间：2013年被农业部列为首批中国重要农业文化遗产。

Geographical location: 41°12′-43°40′N, 87°16′-91°55′E; located in Turpan City, Xinjiang Uygur Autonomous Region.
Climate type: temperate continental arid desert climate.
Time of identification: In 2013, it was listed as one of the first batch of the China-NIAHS by the Ministry of Agriculture.

Large-Scale Underground Agricultural Irrigation Project
Turpan Karez Agricultural System

In Turpan, karezs can be called the "life well".

Turpan is located in the Chinese mainland, with a temperate desert climate that makes it hot and dry. Compared with an average annual evaporation of more than 3,000mm, 16mm of precipitation is almost negligible. Turpan is called "fire land" for its extremely dry climate. Such environment does not seem suitable for human survival and agricultural development, but Turpan is an oasis in the desert.

Turpan is surrounded by mountains, with Ala Mountain in the west, Bogda Mountain in the north and Kurutag Mountain in the south. As temperature rises in summer, the snows on the surrounding mountains begin to thaw, producing large amounts of meltwater. The meltwater permeates underground and brings abundant groundwater to Turpan.

People always have a lot of wisdom in using the bounty of nature. In Turpan, people dug canals along the slope of the ground, which eventually led underground water to farmland and residential buildings. This is the karezs, one of the largest underground irrigation systems around the world. The Turpan Karez Agricultural System is an agricultural water conservancy and production system, which uses a unique way to divert alpine snow water into undercurrent and then irrigate farmlands.

Karez is a kind of unique underground water conservancy project, which utilizes ground slope to introduce groundwater into farmland and dwellings. It consists of four parts: a vertical shaft, a closed conduit, an open channel and a waterlogging dam. People build vertical shafts on the ground, and the bottom of the vertical shaft is connected with the closed conduit. The depth of the well decreases with the decline of the slope of the ground. In addition to collecting water and connecting closed conduits in series, the vertical shaft also has the function of ventilating and sending out the soil and playing the role as a passage way. The closed conduit can transport and collect water. The open channel is the direct water diversion area excavated in the lowland according to the slope descent. The waterlogging dam is the end of the karez. Underground water from the closed conduit flows into the open channel, and finally into the waterlogging dam to be preserved. The whole process completely relies on water self-flow to achieve water diversion. Storing water underground also reduces the loss of water due to high local evaporation. It has excellent water quality for irrigation and human and animal drinking.

Karez was first seen in China in the Han Dynasty, which was also recorded in many other books of the Han Dynasty such as the passage *Waterways in Records of the Historian*. At that time, Xinjiang, Gansu and Shaanxi Province already began to use this kind of mean to exploit groundwater.

However, it is still controversial when it comes to the topics of where did the karez originate and when it was first excavated. Anyway, the karez is eventually extended, utilized and preserved in Turpan until today. Such underground irrigation systems can be seen in several arid regions around the world. At its peak, Turpan has more than 1,000 karezs with an annual flow of 560 million cubic meters and an irrigated area of about 350,000 mu. Under the extremely harsh climate and environmental conditions, the karez brings vitality to its agriculture, feeds generations of people in Turpan and makes Turpan an "oasis" in the desert.

水是人类生命之源。在新疆干旱地区，为适应自然环境，当地人奇迹般地创造出一种特殊的灌溉系统，这就是非常有名的新疆吐鲁番坎儿井农业系统。图为犹如珠链般的坎儿井灌溉了吐鲁番盆地的大片农田。（夏铨 摄）

Water is the source of human life. In the arid areas of Xinjiang, the local people have miraculously created a special irrigation system to adapt to the natural environment—the well-known Turpan Karez Agricultural System, The picture shows the pearl-like chain of karezs irrigating a large area of farmlands in Turpan Basin. Photographed by Xia Quan

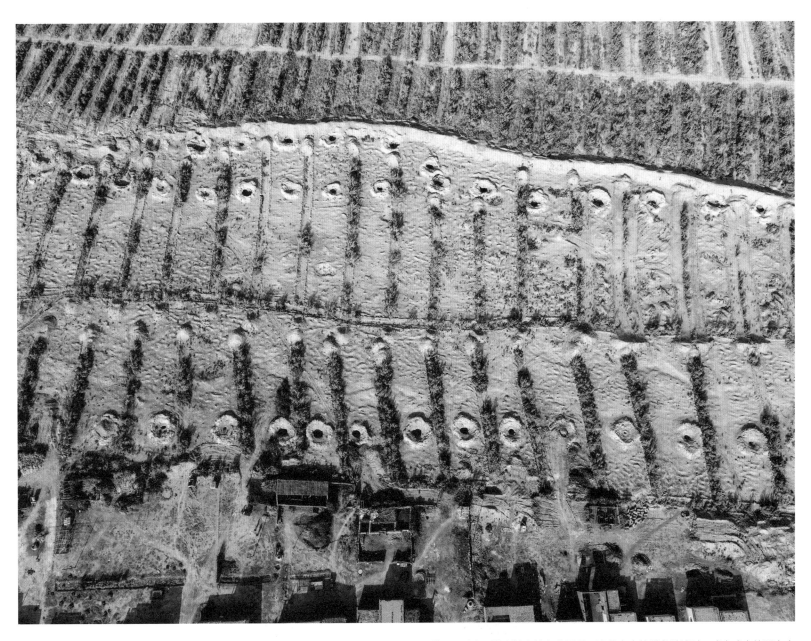

新疆的坎儿井，是吐鲁番绿洲特有的文化景观。当从空中俯瞰戈壁滩时，成串成串的凹心土堆便十分壮观地映入眼帘。坎儿井与农田和民居毗邻，既浇灌了农田，也为当地百姓提供了生活用水。坎儿井被看作中国"干极"的生命之魂，它对发展当地农业生产和满足居民生活需要等都具有很重要的意义。（夏铨 摄）

Xinjiang's karezs are unique cultural landscape in Turpan oasis. Looking down at the Gobi Desert from the air, the clusters of hollow mounds come into view spectacularly. Karez, adjacent to farmland and residential buildings, not only irrigates farmland, but also provides water for local people. The karezs are regarded as the soul of life of "dry pole" of China, which is of great significance to the development of local agricultural production and the satisfaction of residents' living needs. Photographed by Xia Quan

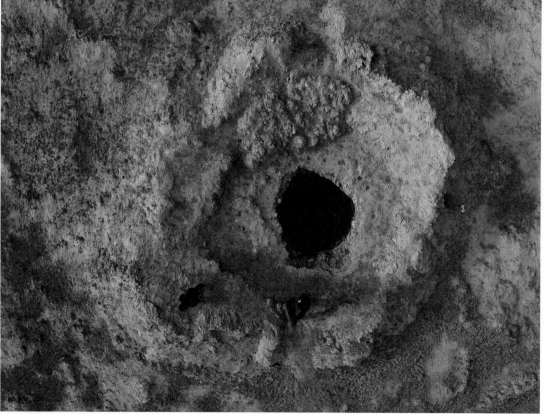

俯瞰坎儿井，古老的农业系统与现代的高压电网构成了人类在这一地区的古今奇观。而从高空俯瞰坎儿井，其中的孔洞即为坎儿井的竖井口，也是通风口。（夏铨 摄）

Overlooking the karezs, the ancient agricultural system and the modern high-voltage power grid constitute a cross-time wonder of human beings in this area. The holes are the well openings and the vents of the karezs. Photographed by Xia Quan

坎儿井名为井，实为人工开挖的地下水渠，主要由竖井、暗渠、明渠、涝坝四部分组成，渠道总长度约5000千米。它是世界上最大的地下水利灌溉系统之一，被誉为"地下万里长城"，是中国古代三大工程（即新疆的坎儿井、万里长城、京杭大运河）之一。图为坎儿井竖井下部联通暗渠，井水从暗渠流到明渠，汇入涝坝，以及当地村民在坎儿井井口洗菜的情景。（夏铨 摄）

Those karezs are actually man-made underground water channels, mainly composed of a vertical shaft, a closed conduit, an open channel and a waterlogging dam. The total length of the channel is about 5,000 km. It is one of the largest underground irrigation system around the world, known as the "Underground Great Wall". It is one of the three greatest ancient Chinese projects (the karezs in Xinjiang, the Great Wall and the Grand Canal from Beijing to Hangzhou). The pictures show that the lower part of the vertical shaft is communicated with the closed conduit, and the well water flows from the closed conduit to the open channel and finally effluxes into the waterlogging dam. It also shows the scene of locals washing vegetables at the opening of the well. Photographed by Xia Quan

如今的坎儿井不少已经失去了灌溉功能，但作为地下水利奇观，有的坎儿井在人们的修整下，具备了观光功能。在一些暗渠内，当地相关部门还加装了照明设备和安全防护设备，供人们切身体会坎儿井的神奇魅力，同时在认识坎儿井的基础上，懂得保护坎儿井，善待坎儿井，使这一伟大农业文化遗产得以继承。（夏铨 摄）

Many karezs today have lost the function of irrigation, but as a wonder of underground water conservancy project, some karezs possess the function of tourism. Local authorities install lighting equipment and safety protection in some of the closed conduits for people to experience the magic charm of karezs. At the same time, the grand agri-cultural heritage can also be well protected and inherited on the basis of understanding it. Photographed by Xia Quan

据调查，吐鲁番坎儿井最多时有1300条左右，年流量5.6亿立方米，灌溉面积约35万亩。吐鲁番著名的葡萄也得益于坎儿井的滋养。上图为当地人在摘葡萄，下图为当地人在晾房中晾晒葡萄干。（夏铨 摄）

According to the survey, there are about 1,300 karezs at its peak in Turpan, with an annual flow of 560 million cubic meters and an irrigation area of about 350,000 mu. The well-known grapes of Turpan are also nourished by karezs. The picture above shows the local people are picking grapes and the below shows they are drying grapes for raisins in a drying room. Photographed by Xia Quan

哈密地域文化与财富的标志

新疆哈密市哈密瓜栽培与贡瓜文化系统

与哈密瓜相比，位于新疆最东端的哈密市对人们来说显得尤为陌生。殊不知，"哈密瓜"的名称正是由"哈密"这个地名得来。哈密是新疆最早开发的地区之一，也是多民族聚居区。汉代时，中央政府就在这片区域行使管理权。

这片深居大陆腹地的沙漠绿洲，四面环山，气候干燥少雨，光热丰富，昼夜温差大，蒸发量大。不过，恰是这种气候环境条件，孕育出了甜度极高、让人念念不忘的香甜瓜果。

在清代之前，哈密生产的瓜果还不叫"哈密瓜"，只称甜瓜。

是的，哈密瓜并不单指一个瓜的品种，而是这一地区生产的瓜果的统称。根据相关文献记载，哈密地区从汉代就开始种植甜瓜，距今已有2000多年的历史。哈密地方品种的甜瓜有124种，其中有43种广为栽培。

据记载，康熙三十七年（1698年），清廷理藩院郎中布尔塞来哈密编设旗队，哈密一世回王额贝都拉多次以清脆香甜、风味独特的哈密甜瓜招待他。次年冬，哈密回王额贝都拉奉旨入京觐见，又精心挑选了100个哈密甜瓜送到京城，康熙皇帝品尝完后，赞不绝口，以地名赐名"哈密瓜"。

自此以后，哈密回王年年向朝廷进贡"哈密瓜"，并划出了专门的贡瓜种植区域，派专人进行种植看护。此后，人们把哈密地区生产的甜瓜都叫作"哈密瓜"。哈密瓜肉质肥厚、甜而不腻，广受人们欢迎。纪晓岚在《阅微草堂笔记》中，曾夸赞"瓜莫盛于哈密"。新疆哈密市哈密瓜栽培与贡瓜文化系统即是以贡瓜为特色的哈密瓜栽培文化系统。

岁岁进贡哈密瓜的传统，直至清光绪年间（1875—1908年）"旋因圣主体恤藩臣，恐途长道远，解运维艰，不依口腹累人"而被废止。哈密瓜作为皇家贡品的身份结束了，但是它的名声却不断传播开来。直至今天，哈密瓜的名头已远盛于当年的丝路重镇哈密，成为哈密文化的一种代表。

哈密瓜是上天赐给哈密这块绿洲的瑰宝，在新丝绸之路经济带上，哈密瓜栽培与贡瓜文化系统作为弥足珍贵的农业文化遗产，仍在为世界增添一抹绿色，为生活增添一缕瓜香。

地理位置：北纬40°52′47″—45°05′33″、东经91°06′33″—96°23′00″，地处新疆维吾尔自治区哈密市。
气候特点：温带大陆性干旱气候。
认定时间：2014年被农业部列为第二批中国重要农业文化遗产。

Geographical location: 40°52′47″-45°05′33″N, 91°06′33″-96°23′00″E; located in Hami City, Xinjiang Uygur Autonomous Region.
Climate type: temperate continental arid climate.
Time of identification: In 2014, it was listed as one of the second batch of the China-NIAHS by the Ministry of Agriculture.

Symbol of Regional Culture and Wealth of Hami City
Hami Melon Cultivation and Tribute Melon Culture System

Compared with Hami melon, Hami City, which is located in the easternmost end of Xinjiang, is particularly strange to us. Little did we know that the name of "Hami melon" was derived from the place named "Hami". Hami is one of the earliest developed areas in Xinjiang and also a multi-ethnic community. During the Han Dynasty, the central government had already exercised its administrative power in this area.

This desert oasis, deep in the hinterland of the mainland and surrounded by mountains, has a dry climate, less rain, abundant light and heat, a large temperature difference between day and night and a large amount of evaporation. However, it is precisely this climate and environmental conditions that breed sweet melons and fruits which are extremely sweet and memorable.

Before the Qing Dynasty, the melons and fruits produced in Hami were not called "Hami melons", only named melons. Yes, Hami melon is not just a variety of melon, but a general term for the melons and fruits produced in this area. According to the relevant documents, melons have been planted in Hami since the Han Dynasty, which has enjoyed a history of more than 2000 years. There are 124 varieties of melons in Hami, 43 of which are widely cultivated.

It is recorded that in the thirty-seventh year of Kangxi (1698), a famous emperor of the Qing Dynasty, Burserai, an official of the Qing Dynasty's Li Fan Court, an institution similar to today's Ministry of Minority Affairs was assigned to set up a flag team in Hami. The king of Hami Erbedura treated him many times with Hami melon, which is crisp, sweet and has a unique flavor. In the winter of the following year, Ebedura went to court in Beijing on imperial order, and carefully selected 100 Hami melons and sent them to the capital. After tasting them, Emperor Kangxi was full of praise and gave them the name "Hami melon".

Since then, Kings of Hami area paid tributes to the court every year with "Hami melons", and designated a special area for the cultivation of tribute melon, sent someone special to plant and care for them. Since then, the melons produced in Hami area have been called "Hami melons". Hami melon is very popular because it's fleshy, sweet but not greasy. Ji Xiaolan, a Qing Dynasty statesman and scholar, praised Hami melon in his work *Yuewei Cottage Notes* "Melons produced in Hami is unequalled". The Hami Melon Cultivation and Tribute Melon Culture System, Xinjiang is the culture system of Hami Melon Cultivation characterized by tribute melons.

The tradition of paying tributes to court with Hami melons was abolished until the reign of Emperor Guangxu in the Qing Dynasty, when "the Lord showed understanding to his vassal for the tough efforts made on such a long way just to meet the emperor's appetite. Hami melon's status as a royal tribute ended, but its reputation continued to spread. Up to now, Hami melon has become a representative of Hami culture, which is far more famous than the Silk Road Township of Hami.

Hami melon is a treasure bestowed by heaven on the oasis of Hami. On the new Silk Road Economic Belt, Hami Melon Cultivation and Tribute Melon Culture System still adds a touch of green to the world and a wisp of melon fragrance to life as a precious agricultural heritage system.

哈密瓜栽培与贡瓜文化系统位于新疆维吾尔自治区哈密市，哈密市地处新疆最东端，是新疆的东大门，自古就是丝绸之路上的重镇，有着悠久的历史和灿烂的文化。图为在新疆哈密市伊州区南湖乡拖布塔村的哈密瓜园。（蔡增乐 摄）

Hami Melon Cultivation and Tribute Melon Culture System located in Hami City, Xinjiang Uygur Autonomous Region, is the eastern gate of Xinjiang, and since ancient times, it has been an important town with a long history and brilliant culture on the Silk Road. The picture shows the Hami melon gardens in Tuobuta Village, Nanhu Township, Yizhou District, Hami City, Xinjiang. Photographed by Cai Zengle

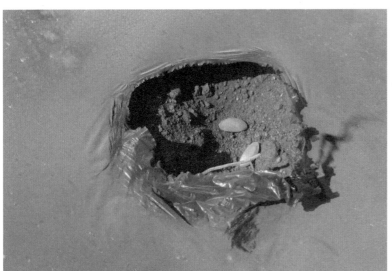

哈密是哈密瓜的故乡，以盛产哈密瓜闻名于世，瓜以地名，地以瓜闻。哈密瓜是在哈密特定的气候条件和自然环境中孕育出来的名优产品，主要种植区域集中在哈密市花园乡、南湖乡等地。图为当地瓜农 3—4 月份在为瓜田施肥、点种和覆膜。（蔡增乐、李华 摄）

Hami is the hometown of Hami melon. It is famous for its abundance of Hami melon all over the world. Hami melon is named after the name of Hami area and Hami area is famous for the melons produced here. Hami melon is a famous and excellent product bred in the specific climate conditions and natural environment of Hami area. The main planting areas are concentrated in Hami's Garden Township, Nanhu Township and other places. The pictures show local melon farmers applying fertilizers, dotting seeds and mulching film in melon fields from March to April. Photographed by Cai Zengle and Li Hua

"哈密瓜"由康熙皇帝赐名，长期作为贡品贡至朝廷，"贡瓜年年渡卢沟"成为定例。为保证贡瓜风味品质，哈密回王曾划出了专门的贡瓜种植基地，指派专人种植，在品种选择、施肥浇水、栽种管理、收获贮运等方面都做了精心安排。图为今天的哈密瓜种植传承人分取哈密瓜种子，年轻的哈密瓜种植传承人手捧哈密瓜苗走向田间。（李华 摄）

"Hami melon" was named by Emperor Kangxi. It used to be a tribute to the court for a long time. It became a set rule that "Tribute melon crossed Lugou River every year". In order to ensure the flavor and quality of tribute melons, Kings of Hami area had designated a special planting base for tribute melons, appointed special people to plant them, and made careful arrangements in many aspects such as variety selection, fertilization and watering, planting management, harvesting, storage and transportation etc. The pictures show that today's inheritors of Hami melon planting divide the seeds of Hami melon, and the young inheritors of Hami melon planting carry the Hami melon seedlings to the field. Photographed by Li Hua

每年6月底7月初，正是哈密瓜成熟的季节，很多客商直接到瓜田选瓜并包装外运，一些精品哈密瓜甚至通过专机外运。（蔡增乐 摄）

At the end of June and the beginning of July every year, it is the ripening season of Hami melon. Many businessmen go directly to the melon fields to select melons and pack them for shipment. Some fine Hami melons are even shipped by special plane. Photographed by Cai Zengle

西北地区·Northwest China/ 新疆维吾尔自治区·Xinjiang Uygur Autonomous Region

为了保护和传承哈密瓜栽培与贡瓜文化系统，哈密市制定出台了相关的保护规划、管理办法和地方标准，修葺了哈密王府，打造了全国唯一的哈密瓜主题公园——哈密瓜园，建造了哈密瓜历史文化馆，并自1993年连续举办了十几届哈密瓜节。图为2017年第十四届哈密瓜节开幕式、展览的哈密瓜品种、哈密瓜园和人们观看维吾尔族车轮舞表演。（李华 摄）

In order to protect and inherit the Hami Melon Cultivation and Tribute Melon Culture System, Hami City has formulated and promulgated relevant protection planning, management methods and local standards, renovated Hami Royal Palace, built the only theme park of Hami melon in the country—Hami Melon Garden, built the Historical and Cultural Museum of Hami Melon, and has held more than a dozen consecutive Hami Melon Festivals since 1993. The pictures show the opening ceremony of the 14th Hami Melon Festival in 2017; varieties of Hami melon and the Hami Tribute Melon Garden on exhibition; people are watching a uyghur wheel dance performance. Photographed by Li Hua

附 录

中国重要农业文化遗产名单

(截至 2019 年 5 月)

农业部(时称)从 2012 年开始分 4 批认定了 91 项中国重要农业文化遗产,填补了我国遗产保护领域的空白,有力地带动了遗产地农民就业增收。

首批 19 个中国重要农业文化遗产名单
(2013 年 5 月公布)

河北宣化城市传统葡萄园

内蒙古敖汉旱作农业系统

辽宁鞍山南果梨栽培系统

辽宁宽甸柱参传统栽培体系

江苏兴化垛田传统农业系统

浙江青田稻鱼共生系统

浙江绍兴会稽山古香榧群

福建福州茉莉花与茶文化系统

福建尤溪联合梯田

江西万年稻作文化系统

湖南新化紫鹊界梯田

云南红河哈尼稻作梯田系统

云南普洱古茶园与茶文化系统

云南漾濞核桃—作物复合系统

贵州从江侗乡稻鱼鸭系统

陕西佳县古枣园

甘肃皋兰什川古梨园

甘肃迭部扎尕那农林牧复合系统

新疆吐鲁番坎儿井农业系统

第二批 20 个中国重要农业文化遗产名单
(2014 年 7 月公布)

天津滨海崔庄古冬枣园

河北宽城传统板栗栽培系统

河北涉县旱作梯田系统

内蒙古阿鲁科尔沁草原游牧系统

浙江杭州西湖龙井茶文化系统

浙江湖州桑基鱼塘系统

浙江庆元香菇文化系统

福建安溪铁观音茶文化系统

江西崇义客家梯田系统

山东夏津黄河故道古桑树群

湖北羊楼洞砖茶文化系统

湖南新晃侗藏红米种植系统

广东潮安凤凰单枞茶文化系统

广西龙胜龙脊梯田农业系统

四川江油辛夷花传统栽培体系

云南广南八宝稻作生态系统

云南剑川稻麦复种系统

甘肃岷县当归种植系统

宁夏灵武长枣种植系统

新疆哈密市哈密瓜栽培与贡瓜文化系统

第三批23个中国重要农业文化遗产名单（2015年10月公布）

北京平谷四座楼麻核桃生产系统

北京京西稻作文化系统

辽宁桓仁京租稻栽培系统

吉林延边苹果梨栽培系统

黑龙江抚远赫哲族鱼文化系统

黑龙江宁安响水稻作文化系统

江苏泰兴银杏栽培系统

浙江仙居杨梅栽培系统

浙江云和梯田农业系统

安徽寿县芍陂（安丰塘）及灌区农业系统

安徽休宁山泉流水养鱼系统

山东枣庄古枣林

山东乐陵枣林复合系统

河南灵宝川塬古枣林

湖北恩施玉露茶文化系统

广西隆安壮族"那文化"稻作文化系统

四川苍溪雪梨栽培系统

四川美姑苦荞栽培系统

贵州花溪古茶树与茶文化系统

云南双江勐库古茶园与茶文化系统

甘肃永登苦水玫瑰农作系统

宁夏中宁枸杞种植系统

新疆奇台旱作农业系统

第四批 29 个中国重要农业文化遗产名单
（2017 年 6 月公布）

河北迁西板栗复合栽培系统

河北兴隆传统山楂栽培系统

山西稷山板枣生产系统

内蒙古伊金霍洛农牧生产系统

吉林柳河山葡萄栽培系统

吉林九台五官屯贡米栽培系统

江苏高邮湖泊湿地农业系统

江苏无锡阳山水蜜桃栽培系统

浙江德清淡水珍珠传统养殖与利用系统

安徽铜陵白姜种植系统

安徽黄山太平猴魁茶文化系统

福建福鼎白茶文化系统

江西南丰蜜橘栽培系统

江西广昌莲作文化系统

山东章丘大葱栽培系统

河南新安传统樱桃种植系统

湖南新田三味辣椒种植系统

湖南花垣子腊贡米复合种养系统

广西恭城月柿栽培系统

海南海口羊山荔枝种植系统

海南琼中山兰稻作文化系统

重庆石柱黄连生产系统

四川盐亭嫘祖蚕桑生产系统

四川名山蒙顶山茶文化系统

云南腾冲槟榔江水牛养殖系统

陕西凤县大红袍花椒栽培系统

陕西蓝田大杏种植系统

宁夏盐池滩羊养殖系统

新疆伊犁察布查尔布哈农业系统

APPENDIX

List of China Nationally Important Agricultural Heritage Systems

(As of May 2019)

Since 2012, Ministry of Agriculture, as Ministry of Agriculture and Rural Affairs then was called, has identified 91 China Nationally Important Agricultural Heritage Systems in four batches, filling the gap in the field of heritage protection in China and effectively promoted the employment and income of farmers in heritage sites.

The first batch (released in May 2013, totally 19)

Urban Agricultural Heritage of Xuanhua Grape Gardens (Hebei)

Aohan Dryland Farming System (Inner Mongolia)

Anshan Nanguo Pear Cultivation System (Liaoning)

Kuandian Pillar Ginseng Cultivation System (Liaoning)

Xinghua Duotian Traditional Agrosystem (Jiangsu)

Qingtian Rice-Fish Culture System (Zhejiang)

Shaoxing Kuaijishan Ancient Chinese Torreya (Zhejiang)

Fuzhou Jasmine and Tea Culture System (Fujian)

Youxi Lianhe Terraces (Fujian)

Wannian Traditional Rice Culture System (Jiangxi)

Xinhua Ziquejie Terraces (Hunan)

Honghe Hani Rice Terraces System (Yunnan)

Pu'er Traditional Tea Agrosystem (Yunnan)

Yangbi Walnut-Crop Complex System (Yunnan)

Congjiang Dong's Rice-Fish-Duck System (Guizhou)

Jiaxian Traditional Chinese Jujube Garden (Shaanxi)

Gaolan Shichuan Ancient Pear Orchard (Gansu)

Diebu Zagana Agriculture-Forestry-Animal Husbandry Composite System (Gansu)

Turpan Karez Agricultural System (Xinjiang)

The Second batch (released in July 2014, totally 20)

Cuizhuang Ancient Winter Jujube Orchard (Tianjin)

Kuancheng Traditional Chestnut Cultivation System (Hebei)

Shexian Dryland Farming Terraces System (Hebei)

Aru Horqin Grassland Nomadic System (Inner Mongolia)

Hangzhou West Lake Longjing Tea Culture System (Zhejiang)

Huzhou Mulberry-dyke & Fish-pond System (Zhejiang)

Qingyuan Shiitake Mushroom Culture System (Zhejiang)

Anxi Tie Guanyin Tea Culture System (Fujian)

Chongyi Hakka Terraces (Jiangxi)

Traditional Mulberry System in Xiajin's Ancient Yellow River Course (Shandong)

Yangloudong Brick Tea Culture System (Hubei)

Xinhuang Red Rice Planting System of Dong Nationality (Hunan)

Chao'an Phoenix Single Cluster Tea Culture System (Guangdong)

Longsheng Longji Terraces (Guangxi)

Jiangyou Traditional Magnolia Cultivation System (Sichuan)

Guangnan Babao Rice Farming Ecosystem (Yunnan)

Jianchuan Rice-Wheat Multiple Cropping System (Yunnan)

Minxian Angelica Planting System (Gansu)

Lingwu Long Jujube Planting System (Ningxia)

Hami Melon Cultivation and Tribute Melon Culture System (Xinjiang)

The third batch (released in October 2015, totally 23)
Pinggu Sizuolou Juglans Hopeiensis Production System (Beijing)
Jingxi Rice Culture System (Beijing)
Huanren Jingzudao Cultivation System (Liaoning)
Yanbian Apple-pear Cultivation System (Jilin)
Fuyuan Fish Culture System of Hezhe Nationality (Heilongjiang)
Ning'an Xiangshui Rice Culture System (Heilongjiang)
Taixing Ginkgo Biloba Cultivation System (Jiangsu)
Xianju Myrica Rubra Cultivation System (Zhejiang)
Yunhe Terrace Agricultural System (Zhejiang)
Shouxian Qubei Irrigation Project (Anfengtang) and Irrigated Agricultural System (Anhui)
Xiuning Spring Water Aquaculture System (Anhui)
Zaozhuang Ancient Jujube Forest (Shandong)
Laoling Jujube Forest Compound System (Shandong)
Lingbao Chuan-Yuan Ancient Jujube Forest (Henan)
Enshi Jade Leaf Tea Culture System (Hubei)
Long'an "Na" Rice Culture System of Zhuang Nationality (Guangxi)
Cangxi Snow-pear Cultivation System (Sichuan)
Meigu Tartary Buckwheat Cultivation System (Sichuan)
Huaxi Ancient Tea Culture System (Guizhou)
Shuangjiang Mengku Ancient Tea Culture System (Yunnan)
Yongdeng Kushui Rose Farming System (Gansu)
Zhongning Lycium Barbarum Planting System (Ningxia)
Qitai Dry Farming System (Xinjiang)

The fourth batch (released in June 2017, totally 29)
Qianxi Chestnut Complex Cultivation System (Hebei)
Xinglong Traditional Hawthorn Cultivation System (Hebei)
Jishan Board Jujube Production System (Shanxi)
Yijin Huoluo Farming and Animal Husbandry System (Inner Mongolia)
Liuhe Vitis Amurensis Cultivation System (Jilin)
Jiutai Wuguantun Tribute Rice Cultivation System (Jilin)
Gaoyou Lake Wetland Agricultural System (Jiangsu)
Wuxi Yangshan Prunus Cultivation System (Jiangsu)
Deqing Traditional Freshwater Pearl Cultivation System (Zhejiang)
Tongling White Ginger Planting System (Anhui)
Huangshan Taiping Kowkui Tea Culture System (Anhui)
Fuding White Tea Culture System (Fujian)
Nanfeng Tangerine Cultivation System (Jiangxi)

Guangchang Lotus Culture System (Jiangxi)

Zhangqiu Green Onion Cultivation System (Shandong)

Xin'an Traditional Cherry Planting System (Henan)

Xintian Sanwei Pepper Planting System (Hunan)

Huayuan Zila Tribute Rice Culture System (Hunan)

Gongcheng Persimmon Cultivation System (Guangxi)

Haikou Yangshan Litchi Planting System (Hainan)

Qiongzhong Shanlan Rice Culture System (Hainan)

Shizhu Coptis Production System (Chongqing)

Yanting Leizu Silkworm Mulberry Production System (Sichuan)

Mingshan Mengding Tea Culture System (Sichuan)

Tengchong Buffalo Breeding System of Penang River (Yunnan)

Fengxian Dahongpao Zanthoxylum Cultivation System (Shaanxi)

Lantian Apricot Planting System (Shaanxi)

Yanchi Tan Sheep Breeding System (Ningxia)

Yili Chabuchaerbuha Agricultural System (Xinjiang)

附　录

中国的全球重要农业文化遗产名单

(截至2018年4月)

按照联合国粮食及农业组织（FAO）的定义，全球重要农业文化遗产是："农村与其所处环境长期协同进化和动态适应下所形成的独特的土地利用系统和农业景观，这种系统与景观具有丰富的生物多样性，而且可以满足当地社会经济与文化发展的需要，有利于促进区域可持续发展。"

自2002年联合国粮农组织发起"全球重要农业文化遗产"倡议以来，中国便成为这项事业的最早响应者、成功实践者和重要推动者。2012年，中国率先在世界上开展国家级农业文化遗产发掘，现已认定中国重要农业文化遗产91项；2014年，中国政府与联合国粮农组织签署了南南合作框架下开展全球重要农业文化遗产工作合作协议；2015年，中国颁布全球首个《重要农业文化遗产管理办法》；自2016年以来，农业文化遗产发掘与保护工作连续被写入中央一号文件。截至2018年4月，中国已有15项全球重要农业文化遗产，数量位居世界首位。

项目名称	入选时间	所在地区
浙江青田稻鱼共生系统	2005 年	浙江省
江西万年稻作文化系统	2010 年	江西省
云南红河哈尼稻作梯田系统	2010 年	云南省
贵州从江侗乡稻鱼鸭系统	2011 年	贵州省
云南普洱古茶园与茶文化系统	2012 年	云南省
内蒙古敖汉旱作农业系统	2012 年	内蒙古自治区
浙江绍兴会稽山古香榧群	2013 年	浙江省
河北宣化城市传统葡萄园	2013 年	河北省
福建福州茉莉花与茶文化系统	2014 年	福建省
江苏兴化垛田传统农业系统	2014 年	江苏省
陕西佳县古枣园	2014 年	陕西省
浙江湖州桑基鱼塘系统	2017 年	浙江省
甘肃迭部扎尕那农林牧复合系统	2017 年	甘肃省
中国南方山地稻作梯田系统（福建尤溪联合梯田、江西崇义客家梯田、湖南新化紫鹊界梯田、广西龙胜龙脊梯田）	2018 年	福建省、江西省、湖南省、广西壮族自治区
山东夏津黄河故道古桑树群	2018 年	山东省

APPENDIX

List of Globally Important Agricultural Heritage Systems of China

(As of April 2018)

According to The Food and Agriculture Organization of the United Nations(FAO), Globally Important Agricultural Heritage Systems (GIAHS) are "Remarkable land use systems and landscapes which are rich in biological diversity evolving from the ingenious and dynamic adaptation of a community /population to its environment and the needs and aspirations for sustainable development."

Since 2002, when the FAO launched the Globally Important Agricultural Heritage Systems Initiative, China has become the earliest responder, successful practitioner and important promoter of this undertaking. In 2012, China took the lead in exploring national-level agricultural heritage systems all over the world, and has now identified 91 China-NIAHS sites. In 2014, the Chinese government signed a cooperation agreement with the FAO to carry out the work of GIAHS under the framework of South-South Cooperation. In 2015, China promulgated the world's first *Measures for the Management of Important Agricultural Heritage Systems*. Since 2016, the work of agri-cultural heritage has been written into the No. 1 document of the Central Committee. As of April 2018, 15 GIAHS sites of China had been designated by FAO, ranking No.1 in the world.

Project Name	Time of	Selection Location
Qingtian Rice-Fish Culture System	2005	Zhejiang Province
Wannian Traditional Rice Culture System	2010	Jiangxi Province
Honghe Hani Rice Terraces System	2010	Yunnan Province
Congjiang Dong's Rice-Fish-Duck System	2011	Guizhou Province
Pu'er Traditional Tea Agrosystem	2012	Yunnan Province
Aohan Dryland Farming System	2012	Inner Mongolia Autonomous Region
Shaoxing Kuaijishan Ancient Chinese Torreya	2013	Zhejiang Province
Urban Agricultural Heritage of Xuanhua Grape Gardens	2013	Hebei Province
Fuzhou Jasmine and Tea Culture System	2014	Fujian Province
Xinghua Duotian Traditional Agrosystem	2014	Jiangsu Province
Jiaxian Traditional Chinese Jujube Garden	2014	Shannxi Province
Huzhou Mulberry-dyke & Fish-pond System	2017	Zhejiang Province
Diebu Zagana Agriculture-Forestry-Animal Husbandry Composite System	2017	Gansu Province
Rice Terraces in Southern Mountainous and Hilly Areas in China (Youxi Lianhe Terraces, Chongyi Hakka Terraces, Xinhua Ziquejie Terraces, Longsheng Longji Terraces)	2018	Fujian Province, Jiangxi Province, Hunan Province and Guangxi Zhuang Autonomous Region
Traditional Mulberry System in Xiajin's Ancient Yellow River Course	2018	Shandong Province

图书在版编目（CIP）数据

中国重要农业文化遗产影像志：汉、英 / 高扬主编. -- 北京：中国摄影出版传媒有限责任公司，2019.8
 ISBN 978-7-5179-0877-7

Ⅰ．①中… Ⅱ．①高… Ⅲ．①农业－文化遗产－中国－摄影集 Ⅳ．①S-64

中国版本图书馆CIP数据核字（2019）第158345号

中国重要农业文化遗产影像志

主　　编：	高　扬
执行主编：	郑丽君
中文撰稿：	闵庆文　王晓蓉
英文翻译：	丁树亭　李　梅　严立梅
责任编辑：	郑丽君
内文设计：	胡佳南
封面设计：	冯　卓
出　　版：	中国摄影出版传媒有限责任公司（中国摄影出版社）
	地址：北京市东城区东四十二条48号　邮编：100007
	发行部：010-65136125　65280977
	网址：www.cpph.com
	邮箱：distribution@cpph.com
印　　刷：	北京科信印刷有限公司
开　　本：	12
印　　张：	34.5
版　　次：	2019年8月第1版
印　　次：	2019年8月第1次印刷

ISBN 978-7-5179-0877-7

定　　价：298.00元

版权所有　侵权必究